Second Edition

Satellite Communication Engineering

Second Edition

Satellite Communication Engineering

Michael Olorunfunmi Kolawole

CRC Press
Taylor & Francis Group
Boca Raton   London   New York

CRC Press is an imprint of the
Taylor & Francis Group, an **informa** business

CRC Press
Taylor & Francis Group
6000 Broken Sound Parkway NW, Suite 300
Boca Raton, FL 33487-2742

First issued in paperback 2017

Version Date: 20130715

ISBN 13: 978-1-4822-1010-1 (hbk)
ISBN 13: 978-1-138-07535-1 (pbk)

Library of Congress Cataloging-in-Publication Data

Kolawole, Michael O.
  Satellite communication engineering / Michael Olorunfunmi Kolawole. -- 2nd edition.
    pages cm
  Includes bibliographical references and index.
    ISBN 978-1-4822-1010-1 (hardback : alk. paper) 1. Artificial satellites in
    telecommunication. I. Title.

TK5104.K55 2013
621.382'5--dc23   2013023566

Visit the Taylor & Francis Web site at
http://www.taylorandfrancis.com

and the CRC Press Web site at
http://www.crcpress.com

Contents

Preface

This second edition of *Satellite Communication Engineering* is an undeniably rich guide to satellite communication principles and engineering with inclusion of recent developments enabling digital information transmission and delivery via satellite.

Satellite communication is one of the most impressive spin-offs from the space programs, and has made a major contribution to the pattern of international communications. The engineering aspect of satellite communications combines such diverse topics as antennas, radiowave propagation, signal processing, data communication, modulation, detection, coding, filtering, orbital mechanics, and electronics. Each of these is a major field of study, and each has its own extensive literature. *Satellite Communication Engineering* emphasizes the relevant material from these areas that is important to the book's subject matter and derives equations that the reader can follow and understand.

The aim of this book is to present in a simple and concise manner the fundamental principles common to the majority of information communications systems. Mastering the basic principles permits moving on to concrete realizations without great difficulty. Throughout, concepts are developed mostly on an intuitive, physical basis, with further insight provided by means of a combination of applications and performance curves. Problem sets are provided for those seeking additional training. Starred sections containing basic mathematical development may be skipped with no loss of continuity by those seeking only a qualitative understanding. The book is addressed to electrical, electronics, and communication engineering students, as well as practicing engineers wishing to familiarize themselves with the broad field of information transmission, particularly satellite communications.

The first of the book's eight chapters covers the basic principles of satellite communications, including message security (cryptology).

Chapter 2 discusses the technical fundamentals for satellite communications services, which do not change rapidly as technology, and provides the reader with the tools necessary for calculation of basic orbit characteristics such as period, dwell time, and coverage area; antenna system specifications such as type, size, beamwidth, and aperture-frequency product; and power system design. The system building blocks comprising satellite transponder and system design procedures are also described. While acknowledging that systems engineering is a discipline on its own, it is my belief that the reader will gain a broad understanding of the system engineering design procedure, accumulated from my experience in large, complex turnkey projects.

Earth station, which forms the vital part of the overall satellite system, is the central theme of Chapter 3. The basic intent of data transmission is to provide quality transfer of information from the source to the receiver with minimum error due to noise in the transmission channel. To ensure quality information requires smart signal processing technique (modulation) and efficient use of system bandwidth (coding, which is discussed extensively in Chapter 6). The most popular forms of modulation employed in digital communications, such as binary phase shift keying (BPSK), quadrature phase shift keying (QPSK), offset-quadrature phase shift keying (OQPSK), and 8-ary phase shift keying (8-PSK), are discussed together with their performance criteria (bit error rate [BER]). An overview of information theory is given to enhance the reader's understanding of how maximum data can be transmitted reliably over the communication medium. Chapter 3 concludes by describing (1) a method for calculating system noise temperature, (2) elements of earth station design, and (3) antenna tracking and the items that facilitate primary terrestrial links to and from the earth stations.

Chapter 4 discusses the process of designing and calculating the carrier-to-noise ratio as a measure of the system performance standard. The quality of signals received by the satellite transponder and that retransmitted and received by the receiving earth station is important if successful information transfer via the satellite is to be achieved. Within constraints of transmitter power and information channel bandwidth, a communication system must be designed to meet certain minimum performance standards. The most important performance standard is the energy bit per noise density in the information channel, which carries the signals in a format in which they are delivered to the end users.

To broadcast video, data, or audio signals over a wide area to many users, a single transmission to the satellite is repeated and received by multiple receivers. While this might be a common application of satellites, there are others that may attempt to exploit the unique capacity of a satellite medium to create an instant network and connectivity between any points within its view. To exploit this geometric advantage, it is necessary to create a system of multiple accesses in which many transmitters can use the same satellite transponder simultaneously. Chapter 5 discusses the sharing techniques called multiple access. Sharing can be in many formats, such as sharing the transponder bandwidth in separate frequency slots (frequency division multiple access [FDMA]), sharing the transponder availability in time slots (time division multiple access [TDMA]), or allowing coded signals to overlap in time and frequency (code division multiple access [CDMA]). The relative performance of these sharing techniques is discussed.

Chapter 6 explores the use of error-correcting codes in a noisy communication environment, and how transmission error can be detected and correction effected using the forward error correction (FEC) methods, namely, the linear block and convolutional coding techniques. Examples are sparingly used as illustrative tools to explain the FEC techniques.

The regulation that covers satellite networks occurs on three levels: international, regional, and national. Chapter 7 discusses the interaction among these three regulatory levels.

Customer's demands for personalized services and mobility, as well as provision of standardized system solutions, have caused the proliferation of telecommunications systems. Chapter 8 examines basic mobile satellite systems services and their interaction with land-based backbone networks—in particular the integrated services digital network (ISDN). Since the services covered by ISDN should also, in principle, be provided by a digital satellite network, it is necessary to discuss in some detail the basic architecture of ISDN as well as its principal functional groups in terms of reference configurations, applications, and protocols. Chapter 8 concludes by briefly looking at the cellular mobile system, including cell assignment and internetworking principles, as well as technological obstacles to providing efficient Internet access over satellite links.

Michael Olorunfunmi Kolawole

Acknowledgments

The inspiration for writing *Satellite Communication Engineering* comes partly from my students who have wanted me to share the wealth of my experience acquired over the years and to ease students' burden in understanding the fundamental principles of satellite communications. Very special thanks go to my darling wife, Dr. Marjorie Helen Kolawole, who actively reminds me about my promise to my students, and more importantly to transfer knowledge to a wider audience. I am eternally grateful for their vision and support.

I also thank Professor Patrick Leung of Victoria University, Melbourne, Australia, for his review of the first edition and for his constructive criticisms, and acknowledge the anonymous reviewers for their helpful comments.

Finally, I thank my family for sparing me the time, which I would have otherwise spent with them, and their unconditional love that keeps me going.

The Author

Dr. Michael O. Kolawole is a distinguished educator and practitioner. He is the director of Jolade Consulting Company, Melbourne, and has held adjunct and visiting professorial appointments in Australia and Nigeria, including the Federal University of Technology, Akure. He has published more than 50 papers in technical journals and 25 industry-based technical reports covering design, performance evaluation, and developmental algorithms. He has overseen a number of operational innovations and holds two patents. He is the author of *Satellite Communication Engineering* (New York: Marcel Dekker, 2002), *Radar Systems, Peak Detection and Tracking* (Oxford: Elsevier, 2003), and *A Course in Telecommunication Engineering* (New Delhi: S Chand, 2009), and co-author of *Basic Electrical Engineering* (Akure: Aoge Publishers, 2012).

Dr. Kolawole received a B.Eng. (1986) degree from Victoria University, Melbourne, and Ph.D. (2000) degree from the University of New South Wales, Sydney, both in electrical engineering. He also received an M.S. (1989) degree from the University of Adelaide in environmental studies. Dr. Kolawole is a chartered professional engineer in Australia, and a member of the New York Academy of Sciences. He plays clarinet and saxophone, and composes and arranges music.

1

Basic Principles of Satellite Communications

Satellite communication is one of the most impressive spin-offs from the space programs, and has made a major contribution to the pattern of international communications. A communication satellite is basically an electronic communication package placed in orbit whose prime objective is to initiate or assist communication transmission of information or messages from one point to another through space. The information transferred most often corresponds to voice (telephone), video (television), and digital data.

Communication involves the transfer of information between a source and a user. An obvious example of information transfer is through terrestrial media, through the use of wire lines, coaxial cables, optical fibers, or a combination of these media.

Communication satellites may involve other important communication subsystems as well. In this instance, the satellites need to be monitored for position location in order to instantaneously return an upwardly transmitting (uplink) ranging waveform for tracking from an earth terminal (or station). The term *earth terminal* refers collectively to the terrestrial equipment complex concerned with transmitting signals to and receiving signals from the satellite. The earth terminal configurations vary widely with various types of systems and terminal sizes. An earth terminal can be fixed and mobile land based, sea based, or airborne. Fixed terminals, used in military and commercial systems, are large and may incorporate network control center functions. Transportable terminals are movable but are intended to operate from a fixed location, that is, a spot that does not move. Mobile terminals operate while in motion; examples are those on commercial and navy ships as well as those on aircraft. Chapter 3 addresses a basic earth terminal configuration.

Vast literature has been published on the subject of satellite communications. However, the available literature appears to deal specifically with specialized topics related to communications techniques, design or part thereof, or satellite systems as a whole.

This chapter looks briefly at the development and principles of satellite communication and its characteristic features.

1.1 The Origin of Satellites

The space age began in 1957 with the USSR's launch of the first artificial satellite, called *Sputnik*, which transmitted telemetry information for 21 days. This achievement was followed in 1958 by the American artificial satellite, *Score*, which was used to broadcast President Eisenhower's Christmas message. Two satellites were deployed in 1960: a reflector satellite, called *Echo*, and *Courier*. The *Courier* was particularly significant because it recorded a message that could be played back later. In 1962 active communication satellites (repeaters), called *Telstar* and *Relay*, were deployed, and the first geostationary satellite, called *Syncom*, was launched in 1963. The race for space exploitation for commercial and civil purposes truly started.

A satellite is *geostationary* if it remains relatively fixed (stationary) in an apparent position relative to the earth. This position is typically about 35,784 km away from the earth. Its elevation angle is orthogonal (i.e., 90°) to the equator, and its period of revolution is synchronized with that of the earth in inertial space. A geostationary satellite has also been called a *geosynchronous* or *synchronous* orbit, or simply *geosatellite*.

The first series of commercial geostationary satellites (*Intelsat* and *Molnya*) was inaugurated in 1965. These satellites provided video (television) and voice (telephone) communications for their audiences. Intelsat was the first commercial global satellite system owned and operated by a consortium of more than 100 nations, hence its name, which stands for *International Telecommunications Satellite Organization*. The first organization to provide global satellite coverage and connectivity, it continues to be the major communications provider with the broadest reach and the most comprehensive range of services.

Other providers for industrial and domestic markets include *Westar* in 1974, *Satcom* in 1975, *Comstar* in 1976, *SBS* in 1980, *Galaxy* and *Telstar* in 1983, *Spacenet* and *Anik* in 1984, *Gstar* in 1985, *Aussat* in 1985–1986, *Optus A2* in 1985, *Hughes-Ku* in 1987, *NASA ACTS* in 1993, *Optus A3* in 1997, *Iridium* and *Intelsat VIIIA* in 1998, and *Optus C* in 2003. With technology for communication satellites improving with time, competition among nations has increased to provide and meet domestic communication as well as security needs, for example, *Sina-1* in 2005, *NigeriaSat-2* and *NigeriaSat-X* in 2011, *Zhongxing-1A* in 2011, *GSAT-10* in 2012, *Zenit-3SL* in 2006, and *SATCOM-4* in 2012. Even more are planned. Some of these satellites host dedicated military communication channels. The need to have market domination and a competitive edge in military surveillance and tactical fields results in more sophisticated developments in the satellite field.

1.2 Communications via Satellite

Radiowaves, suitable as carriers of information with a large bandwidth, are found in frequency ranges where the electromagnetic waves are propagated through space almost in conformity with the law of optics, so that only line-of-sight radiocommunication is possible [1]. As a result, topographical conditions and the curvature of the earth limit the length of the radio path. Relay stations, or repeaters, must be inserted to allow the bridging of greater distances (see Figure 1.1). Skywave radar uses the ionosphere, at a height of 70 to 300 km, to transmit information beyond the horizon and may not require repeaters. However, transmission suffers from ionospheric distortions and fading [16]. To ensure that appropriate frequencies are optimally selected, additional monitoring equipment is required to sample the ionospheric conditions instantaneously.

A communication satellite in orbit around the earth exceeds the latter requirement. Depending on the orbit's diameter, satellites can span large distances almost half the earth's circumference. However, a communication link between two subsystems—for instance, earth stations or terminals—via

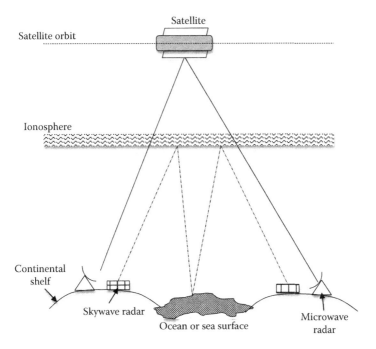

FIGURE 1.1
Intercontinental communication paths.

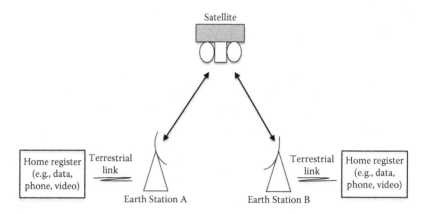

FIGURE 1.2
Communication between two earth stations via a satellite.

the satellite may be considered a special case of radio relay, as shown in Figure 1.2, with a number of favorable characteristics:

A desired link between two terminals in the illumination zone can be established.

The investment for a link in the illumination zone is independent of the distance between the terminals.

A provision for wide-area coverage for remote or inaccessible territories or for new services is made.

This is ideally suited to medium, point-to-multiunit (broadcast) operations.

A practical satellite comprises several individual chains of equipment called a *transponder*—a term derived from *transmitter* and *responder*. Transponders can channel the satellite capacity both in frequency and in power. A transponder may be accessed by one or by several carriers. Transponders exhibit strong nonlinear characteristics and multicarrier operations, unless properly balanced, which may result in unacceptable interference. The structure and operation of a transponder are addressed in Chapter 2, and the techniques used to access the transponder are examined in Chapter 5.

1.3 Characteristic Features of Communication Satellites

Satellite communication circuits have several characteristic features. These include:

1. Circuits that traverse essentially the same radiofrequency (RF) pathlength regardless of the terrestrial distance between the terminals.

2. Circuits positioned in geosynchronous orbits may suffer a transmission delay, t_d, of about 119 ms between an earth terminal and the satellite, resulting in a user-to-user delay of 238 ms and an echo delay of 476 ms.

   For completeness, transmission delay is calculated using

   $$t_d = \frac{h_0}{c} \tag{1.1}$$

   where h_0 is the altitude above the subsatellite point on the earth terminal and c is the speed of light ($c = 3 \times 10^8$ m/s).

   For example, consider a geostationary satellite whose altitude h_0 above the subsatellite point on the equator is 35,784 km. This gives a one-way transmission delay of 119 ms, or a round-trip transmission delay of 238 ms. It should be noted that an earth terminal not located at the subsatellite point would have greater transmission delays.

3. Satellite circuits in a common coverage area pass through a single RF repeater for each satellite link (more is said of the coverage area, repeater, and satellite links in Chapters 2 and 4). This ensures that earth terminals, which are positioned at any suitable location within the coverage area, are illuminated by the satellite antenna(s). The terminal equipment could be fixed or mobile on land or mobile on ship and aircraft.

4. Although the uplink power level is generally high, the signal strength or power level of the received downlink signal is considerably low because of:

   High signal attenuation due to free-space loss

   Limited available downlink power

   Finite satellite downlink antenna gain, which is dictated by the required coverage area

   For these reasons, the earth terminal receivers must be designed to work at significantly low RF signal levels. This leads to the use of the largest antennas possible for a given type of earth terminal (discussed in Chapter 3) and the provision of low-noise amplifiers (LNAs) located in close proximity to the antenna feed.

5. Messages transmitted via the circuits are to be secured, rendering them inaccessible to unauthorized users of the system. Message security is a commerce closely monitored by the security system designers and users alike. For example, Pretty Good Privacy (PGP), invented by Philip Zimmerman, is an effective encryption tool [2]. The U.S. government sued Zimmerman for releasing PGP to the public, alleging that making PGP available to enemies of the United States could endanger national security. Although the lawsuit was later dropped, the use of PGP in many other countries is still illegal.

1.4 Message Security

Customers' (private and government) increasing demand to protect satellite message transmission against passive eavesdropping or active tampering has prompted system designers to make encryption an essential part of satellite communications system design. Message security can be provided through cryptographic techniques. *Cryptology* is the theory of *cryptography* (i.e., the art of writing in or deciphering secret code) and *cryptanalysis* (i.e., the art of interpreting or uncovering the deciphered codes without the sender's consent or authorization).

Cryptology is an area of special difficulty for readers and students because many good techniques and analyses are available but remain the property of the organizations whose main business is secrecy. As such, we discuss the fundamental technique of cryptography in this section without making any specific recommendation.

1.4.1 Basic Cryptographic Functions

Basic cryptography comprises encryption, decryption, and key management unit, as shown in Figure 1.3. *Encryption* (enciphering) is the process of converting messages, information, or data into a form unreadable by anyone except the intended recipient. The encrypted (enciphered) text is called a *cryptogram*. Encrypted data must be deciphered (unlocked, or decrypted) before the recipient can read it. *Decryption* is the unlocking of the locked message—that is, the reverse of encryption. *Key* simply means "password." *Key management* refers to the generation, distribution, recognition, and reception of the cryptographic keys. Cryptographic key management is the most important element of any cryptographic system (simply called cryptosystem) design.

Encryption uses a special system called an *algorithm* to convert the text of the original message (plaintext) into an encrypted form of the message (ciphertext or cryptogram). Algorithms are step-by-step procedures for solving problems in the case of encryption, for enciphering and deciphering a

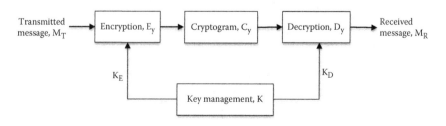

FIGURE 1.3
General cryptographic functions.

plaintext message. Cryptographic algorithms (like *key* and *transformation functions*) equate individual characters in the plaintext with one or more different keys, numbers, or strings of characters. In Figure 1.3, the encryption algorithm E_y transforms the transmitted message M_T into a cryptogram C_y by the cryptographic key K_E algorithm. The received message M_R is obtained through the decryption algorithm D_y with the corresponding decryption key K_D algorithm. These cryptographic functions are concisely written as follows.

Encryption:

$$C_y = E_y(M_T, K_E) \tag{1.2}$$

Decryption:

$$M_R = D_y(C_y, K_D)$$
$$= D_y(E_y(M_T, K_E), K_D) \tag{1.3}$$

1.4.2 Ciphering in Satellite Communication Systems

New-generation equipment aboard satellites and earth stations is extraordinarily sophisticated. It acts as a series of massive electronic vacuum cleaners, sweeping up every kind of communication with embedded security devices. Message security in a satellite network can be achieved by placing the ciphering equipment either at the earth station or in the satellite.

Figure 1.4 shows the block diagrams of messages M_{Ti} transmitted between earth stations via the satellite. Each earth station has a number of cryptographic keys K_{ti} (where $i = 1, 2, ..., n$) shared between the communicating earth stations while the satellite remains transparent, meaning the satellite has no role in the ciphering process.

A cryptosystem that may work for the scenario depicted by Figure 1.4 is described as follows. All the earth stations $TS(i)$ and $RS(j)$ are assumed capable of generating random numbers $RDN(i)$ and $RDN(j)$, respectively. Each transmitting earth station $TS(i)$ generates and stores the random numbers $RDN(i)$. It then encrypts $RDN(i)$, that is, $E_y[RDN(i)]$, and transmits the encrypted random number $E_y[RDN(i)]$ to $RS(j)$. The receiving earth station $RS(j)$ generates $RDN(j)$ and performs a modulo-2 (simply, mod-2) addition with $RDN(i)$, that is, $RDN(j) \oplus RDN(i)$, to obtain the session key $K_r(j)$, where $\oplus$ denotes mod-2 addition. It should be noted that mod-2 addition is implemented with exclusive-OR gates and obeys the ordinary rules of addition, except that $1 \oplus 1 = 0$.

The transmitting earth station retrieves $RDN(i)$ and performs mod-2 addition with $RDN(j)$, that is, $RDN(i) \oplus RDN(j)$, to obtain the session key $K_t(i)$. This process is reversed if $RS(j)$ transmits messages and $TS(i)$ receives.

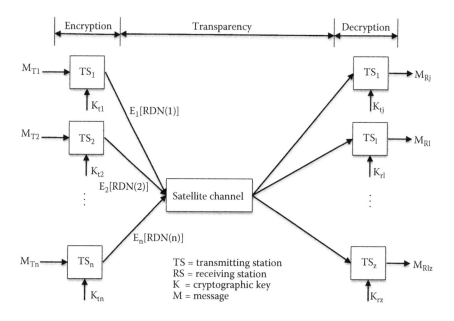

FIGURE 1.4
Earth station-to-earth station ciphering.

Figure 1.5 demonstrates the case where the satellite plays an active ciphering role. In it the keys K_{Ei} (where $i = 1, 2, ..., n$) the satellite receives from the transmitting uplink stations T_{Si} are recognized by the onboard processor, which in turn arranges, ciphers, and distributes to the downlink earth stations, $RS(j)$ (more is said about onboard processing in Chapter 2). Each receiving earth station has matching cryptographic keys K_{Dj} (where $j = l, m, ..., z$) to be able to decipher the received messages M_{Rj}.

The cryptosystem that may work for the scenario depicted by Figure 1.5 is described as follows. It is assumed that the satellite onboard processor is capable of working the cryptographic procedures of the satellite network. It is also assumed that all the earth stations $TS(i)$ and $RS(j)$ play passive roles and only respond to the requests of the satellite onboard processor. The key session of the onboard processor is encrypted under the station master key. The onboard processor's cryptographic procedure provides the key session for recognizing the key session of each earth station. Thus, when an earth station receives encrypted messages from the satellite, the earth station's master key is retrieved from storage. Using the relevant working key to recognize the session key activates the decryption procedure. The earth station is ready to retrieve the original (plaintext) message using the recognized session key.

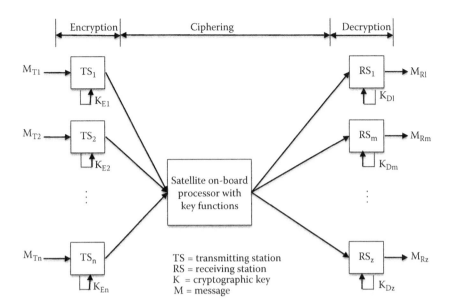

FIGURE 1.5
Ciphering with keyed satellite onboard processor.

The European Telecommunication Satellite (EUTELSAT) has implemented encryption algorithms, for example, the *Data Encryption Standard* (DES), as a way of providing security for their satellite link (more is said about DES in Section 1.4.3).

Having discussed the session key functions K, the next item to discuss is the basic functionality of ciphering techniques and transformation functions.

1.4.3 Ciphering Techniques

Two basic ciphering techniques fundamental to secret system design are discussed in this section: *block ciphering* and *feedback ciphering*.

1.4.3.1 Block Ciphering

Block ciphering is a process by which messages are encrypted and decrypted in blocks of information digits. Block ciphering has the same fundamental structure as block coding for error correction (block coding is further discussed in Chapter 6). Comparatively, a ciphering system consists of an encipher and a decipher, while a coding system consists of an encoder and a decoder. The major difference between the two systems (ciphering and block coding) is that block ciphering is achieved by ciphering keys while coding

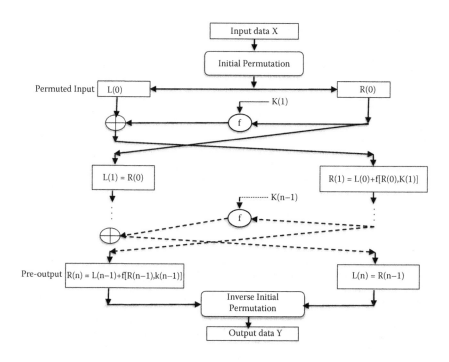

FIGURE 1.6
Block ciphering technique with partition and iteration.

relies on parity checking. A generalized description of the block ciphering technique is shown in Figure 1.6. In block ciphering, system security is
achieved by:

1. Partitioning the message into subblocks, then encrypting (e.g., by
   simple bit permutation and bit inversion) and decrypting (i.e., the
   reverse of encryption) each subblock separately.
2. Repeating the encryption procedure several times. Often in practice,
   the ensuing pattern may be asymmetric, making it difficult for the
   cryptanalyst to break.
3. Combining parts 1 and 2.

The security system designer might use a combination of these procedures to ensure a reasonably secured transmission channel. In 1977 the
U.S. government adopted the preceding partition and iteration procedure as the DES for use in unclassified applications [3]. As of this writing, a new encryption system called the Advanced Encryption Standard
(AES)—a block cipher standardization process—is being revalidated by
the U.S. National Institute of Standards and Technology [4]. AES replaced
the DES, as it uses a more complex algorithm based on a 256-bit *key K*

encryption standard instead of the DES 64-bit standards. The European Electronic Signature Standardization Initiative [5] also adopted the AES. GOST—another encryption system and the official encryption standard of the Russian Federation—uses a 256-bit *key K* block cipher and encryption standardization process [6]. GOST is an acronym for *gosudarstvennyy standart*, which means *state standard*.

The DES system uses *public-key* encryption [3, 4]. In a DES system, each person gets two keys: a public key and a private key. The keys allow a person to either lock (encrypt) a message or unlock (decipher) an enciphered message. Each person's public key is published, and the private key is kept secret. Messages are encrypted using the intended recipient's public key and can only be decrypted using the private key, which is never shared. It is virtually impossible to determine the private key even if you know the public key. In addition to encryption, public-key cryptography can be used for authentication, that is, providing a digital signature that proves a sender or the identity of the recipient. There are other public-key cryptosystems, such as trapdoor [7], the Rivest-Shamir-Adleman (RSA) system [8], McEliece's system [9], elliptic curve cryptography [10, 11], discrete logarithm cryptosystems [12], quantum cryptography [13], and keys with auxiliary inputs [14].

The basic algorithm for block ciphering is shown in Figure 1.6 and described as follows. Suppose there are $n + 1$ iterations to be performed. Denote the input and output data by $X = x_1, x_2, x_3, \ldots, x_m$ and $Y = y_1, y_2, y_3, \ldots, y_m$, respectively. Since the input data to be transformed iteratively is $n + 1$ times, the block of data is divided equally into left and right halves, denoted by $L(j)$ and $R(j)$, respectively, where $j = 0, 1, 2, \ldots, n$. The key of the jth iteration is denoted by $K(j)$. The symbol f denotes a transformation function. There are many processes within Figure 1.6 that require further clarification. The next few subsections attempt to explain these processes, complemented with examples. If we take a segment of the arrangement in Figure 1.6 and reproduce it as Figure 1.7, where the main functions (i.e., encryption, keying, and decryption) are clearly identified, we can write the iteration $j + 1$ from the jth iteration for the encryption function as

$$L(j+1) = R(j)$$

$$R(j+1) = L(j) \oplus f[K(j+1), R(j)] \tag{1.4}$$

Also, we can write the decryption function as

$$L(j) = R(j+1)$$

$$R(j) = L(j+1) \oplus f[K(j+1), R(j+1)] \tag{1.5}$$

where $\oplus$ denotes mod-2 addition.

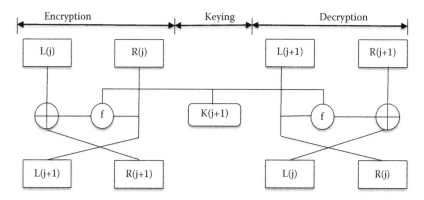

f = transformation function
L(j) = left half of message j
R(j) = right half of message j
K(j+1) = set of iteration dependent keys

FIGURE 1.7
Part of block ciphering with partition and iteration.

1.4.3.1.1 Transformation Function

The transformation function $f[K(j+1), R(j)]$ consists of bit expansion, key mod-2 addition, and selection (or substitution) and permutation operations. Figure 1.8 shows the processes involved in transforming a block of data $R(j)$. The transformation function's operation is a bit mathematically involved. Instead of mathematical representation, each of the functions comprising Figure 1.8 is discussed separately with numerical examples.

1.4.3.1.2 Bit Expansion Function

The function of the bit expansion function is to convert an n-bit block into an $(n + p)$-bit block in accordance with the ordering sequence E_f. The ordering sequence E_f assigns and expands the n-bit block into an extended $(n + p)$-bit block E_x; that is,

$$E_x(j) = E_f[R(j)] \qquad (1.6)$$

The extended function E_x must match the number of bits of the key function $K(j + 1)$.

Example 1.1

As an illustration, define $R(j)$ as a 32-bit block given by

$$R(j) = 10011001110110110101010000111101 \qquad (1.7a)$$

The bit expansion function can be solved by partitioning $R(j)$ into eight segments (columns) with 4 bits in each segment. Ensure that each of the

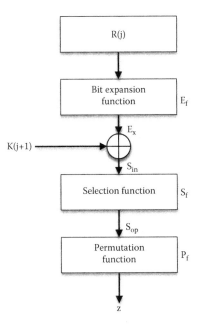

FIGURE 1.8
Sequence of transformation function.

end bits of the segment is assigned to two positions, with the exception of the first and last bits, thus ensuring the ordering sequence of a 48-bit block:

$$
E_f = \begin{matrix}
32 & 1 & 2 & 3 & 4 & 5 \\
4 & 5 & 6 & 7 & 8 & 9 \\
8 & 9 & 10 & 11 & 12 & 13 \\
12 & 13 & 14 & 15 & 16 & 17 \\
16 & 17 & 18 & 19 & 20 & 21 \\
20 & 21 & 22 & 23 & 24 & 25 \\
24 & 25 & 26 & 27 & 28 & 29 \\
28 & 29 & 30 & 31 & 32 & 1
\end{matrix}
\tag{1.7b}
$$

Based on the ordering sequence of (1.7b), the bit expansion function of (1.6) becomes

$$
E_x(j) = E_f[R(j)] = \begin{pmatrix}
1 & 1 & 0 & 0 & 1 & 1 \\
1 & 1 & 0 & 0 & 1 & 1 \\
1 & 1 & 0 & 1 & 1 & 0 \\
1 & 0 & 1 & 1 & 0 & 1 \\
0 & 1 & 0 & 1 & 0 & 1 \\
0 & 1 & 0 & 0 & 0 & 0 \\
0 & 0 & 1 & 1 & 1 & 1 \\
1 & 1 & 0 & 1 & 1 & 0
\end{pmatrix}
\tag{1.8}
$$

which equates to 48 bits, matching the number of bits of key function $K(j+1)$.

1.4.3.1.3 Selection Function

For simplicity, consider an 8-bit selection function S_{fj}, where $j = 1, 2, ...,$ 8. Each function has l rows and $(m + 1)$ columns. The elements of selection function S_{fj} have a specific arrangement of any set of integers from 0 to m. From Example 1.1, each function S_{fj} takes a 6-bit block as its input, S_{in}, denoted by

$$S_{in} = x_1, x_2, x_3, x_4, x_5, x_6 \tag{1.9}$$

Suppose r_j and c_j correspond to a particular row and column of selection function S_{fj}. Row r_j is determined by the first and last digits of S_{in}; that is, (x_1, x_6). Since x_1 and x_6 are binary digits, it follows that there can be only four possible outcomes to indicate the four rows $(l = 4)$ of S_{fj}, while the other four binary digits (x_2, x_3, x_4, x_5) will provide numbers between 0 and m, thereby determining c_j. The intersection of r_j and c_j in S_{fj} produces a specific integer between 0 and m, which, when converted to its binary digits, gives the output

$$S_{op} = y_1, y_2, y_3, y_4 \tag{1.10}$$

In practice, the elements of the selection functions S_{fj} are tabulated like a lookup table. In the case of DES, $l = 4$ and $m = 15$; these elements are shown in Table 1.1. As seen in Table 1.1, each selection function S_{fj} has 4-bit output, giving a total of 32 bits for all selection functions. Example 1.2 demonstrates how the selection function's algorithm is implemented.

Example 1.2

Take a row from (1.8), say the first row, as input data; that is, $S_{in} = 110011$. It follows that $r_j = (x_1, x_6) = 11$ and $c_j = 1001 = 9$. Our task now is to provide the output S_{op} due to S_{in} using the previous transformation process on the basis of the selection functions given by Table 1.1. If we let $j = 8$, the element of S_8 in the fourth row and ninth column is 15, which equates to the digital output $S_{op} = y_1, y_2, y_3, y_4 = 1111$.

1.4.3.1.4 Permutation Function

The purpose of permutation function P_f of Figure 1.8 is to take all the selection function's 32 bits and permute the digits to produce a 32-bit block output. The permutation function P_f simply performs

$$Z = P_f(Y) \tag{1.11}$$

TABLE 1.1

DES Selection Functions

							S1								
14	4	13	1	2	15	11	8	3	10	6	12	5	9	0	7
0	15	7	4	14	2	13	1	10	6	12	11	9	5	3	8
4	1	14	8	13	6	2	11	15	12	9	7	3	10	5	0
15	12	8	2	4	9	1	7	5	11	3	14	10	0	6	13
							S2								
15	1	8	14	6	11	3	4	9	7	2	13	12	0	5	10
3	13	4	7	15	2	8	14	12	0	1	10	6	9	11	5
0	14	7	11	10	4	13	1	5	8	12	6	9	3	2	15
13	8	10	1	3	15	4	2	11	6	7	12	0	5	14	9
							S3								
10	0	9	14	6	3	15	5	1	13	12	7	11	4	2	8
13	7	0	9	3	4	6	10	2	8	5	14	12	11	15	1
13	6	4	9	8	15	3	0	11	1	2	12	5	10	14	7
1	10	13	0	6	9	8	7	4	15	14	3	11	5	2	12
							S4								
7	13	14	3	0	6	9	10	1	2	8	5	11	12	4	15
13	8	11	5	6	15	0	3	4	7	2	12	1	10	14	9
10	6	9	0	12	11	7	13	15	1	3	14	5	2	8	4
3	15	0	6	10	1	13	8	9	4	5	11	12	7	2	14
							S5								
2	12	4	1	7	10	11	6	8	5	3	15	13	0	14	9
14	11	2	12	4	7	13	1	5	0	15	10	3	9	8	6
4	2	1	11	10	13	7	8	15	9	12	5	6	3	0	14
11	8	12	7	1	14	2	13	6	15	0	9	10	4	5	3
							S6								
12	1	10	15	9	2	6	8	0	13	3	4	14	7	5	11
10	15	4	2	7	12	9	5	6	1	13	14	0	11	3	8
9	14	15	5	2	8	12	3	7	0	4	10	1	13	11	6
4	3	2	12	9	5	15	10	11	14	1	7	6	0	8	13
							S7								
4	11	2	14	15	0	8	13	3	12	9	7	5	10	6	1
13	0	11	7	4	9	1	10	14	3	5	12	2	15	8	6
1	4	11	13	12	3	7	14	10	15	6	8	0	5	9	2
6	11	13	8	1	4	10	7	9	5	0	15	14	2	3	12
							S8								
13	2	8	4	6	15	11	1	10	9	3	14	5	0	12	7
1	15	13	8	10	3	7	4	12	5	6	11	0	14	9	2
7	1	4	1	9	12	14	2	0	6	10	13	15	3	5	8
2	1	14	7	4	10	8	13	15	12	9	0	3	5	6	11

where the permutation is like the ordering sequence of the expansion function:

$$P_f = \begin{matrix} 16 & 7 & 20 & 21 \\ 29 & 12 & 28 & 17 \\ 1 & 15 & 23 & 26 \\ 5 & 18 & 31 & 10 \\ 2 & 8 & 24 & 14 \\ 32 & 27 & 3 & 9 \\ 19 & 13 & 30 & 6 \\ 22 & 11 & 4 & 25 \end{matrix} \qquad (1.12a)$$

and the 32-bit block input is

$$Y = y_1, y_2, y_3, \ldots, y_{32} \qquad (1.12b)$$

Hence, on the basis of (1.12), the permutation function can be written as

$$Z = P_f(Y) = y_{16}, y_7, y_{20}, y_{21}, y_{29}, \ldots, y_4, y_{25} \qquad (1.13)$$

which suggests that the 32-bit block input (1.12b) is rearranged (permuted) according to the ordered permutation function given by (1.12a).

Example 1.3

Suppose

$$Y = 10011001110110110101010000111101$$

If we rearrange Y according to (1.12a), the output of the permutation function is

$$Z = 10101110110011010100110101101010$$

In summary, Section 1.4.3.1.2 has demonstrated the process by which an input block data $R(j)$ is transformed to produce an output function, $z(j)$.

1.4.3.2 Feedback Ciphering

A general feedback arrangement is shown in Figure 1.9. Within the mappers G contain the ciphering and deciphering algorithms. The mapping of keys $G(K_i)$ is used for encryption, while $G(K_j)$ is used for decryption.

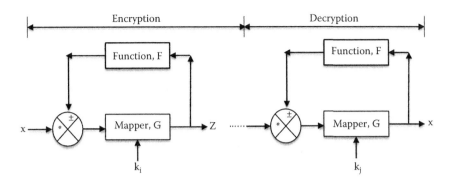

FIGURE 1.9
Feedback ciphering.

For an uncoded message x and the feedback function F, the encrypted message z can be written as

$$z = [x \pm zF]G(K_i) \tag{1.14}$$

Rearranging (1.14) in terms of z, we have

$$z[1 \mp FG(K_i)] = xG(K_i) \tag{1.15a}$$

or

$$z = \frac{G(K_i)}{1 \mp FG(K_i)} x \tag{1.15b}$$

For the decryption circuit, the expression for the decrypted message x can be written as

$$x = [z \pm xF]G(K_j) \tag{1.16a}$$

Alternately, in terms of encrypted message z,

$$z = \frac{1 \mp FG(K_j)}{G(K_j)} x \tag{1.16b}$$

For the same transmission sequence, (1.15b) and (1.16b) must be equal; that is,

$$\frac{1 \mp FG(K_j)}{G(K_j)} = \frac{G(K_i)}{1 \mp FG(K_i)} \tag{1.17}$$

which turns to

$$G(K_i) = 1 \mp FG(K_j)$$
$$G(K_j) = 1 \mp FG(K_i) \tag{1.18}$$

It can be seen in (1.18) that the key functions relate to the feedback functions. If for argument's sake we let the feedback F equate to unity, then the key functions become an additive inverse of each other. This shows that the feedback function and the mapping function are linear functions in this type of ciphering. In general, a feedback mechanism enhances the strength of a cryptosystem [15].

1.5 Summary

This chapter has briefly introduced the genesis and characteristic features of communication satellites. A communication satellite is basically an electronic communication package placed in orbit whose prime objective is to initiate or assist communication transmission of information or messages from one point to another through space. The information being transferred most often corresponds to voice (telephone), video (television), and digital data.

As electronic forms of communication, commerce, and information storage and processing have developed, the opportunities to intercept and read confidential information have grown, and the need for sophisticated encryption has increased. This chapter has also explained basic cryptographic techniques. Many newer cryptography techniques being introduced to the market are highly complex and nearly unbreakable, but their designers and users alike carefully guard their secrets.

The use of satellites for communication has been steadily increasing, and more frontiers will be broken as advances in technology make system production costs economical.

Problems

1. The role of telecommunications networks has changed over the last decade.

    a. Discuss the role of telecommunications networks in modern society.

    b. How has this changed your perception in terms of security, social cohesion, and commerce?

2. Your task is to develop a communication network covering some inhospitable terrain. What sort of telecommunication infrastructure would you suggest? Discuss the social, legal, and political implications of the recommended telecommunications network(s).

3. A packet-switched network is to be designed with onboard processing capability. Design a suitable cryptosystem for securing the information flow and message contents.

References

1. Dressler, W. (1987). Satellite communications. In *Siemens Telecom Report*, vol. 10.
2. Meyer, C., and Matyas, S. (1982). *Cryptography: a new dimension in computer data security*. New York: John Wiley.
3. National Bureau of Standards. (1977). *Data encryption standard*. Federal Information Processing Standards, Publication 46. U.S. Department of Commerce.
4. National Institute of Standards and Technology. (2012). *Advanced encryption standard (AES)*. Federal Information Processing Standards, Publication 197. U.S. Department of Commerce.
5. New European Schemes for Signatures, Integrity, and Encryption (NESSIE). (2004). http://www.cryptonessie.org.
6. Poschmann, A., Ling, S., and Wang, H. (2010). 256 bit standardized crypto for 650 GE GOST revisited. In *Lecture Notes in Computer Science*, LNCS 6225, 219–233.
7. Diffie, W., and Hellman, M. (1976). New directions in cryptography. *IEEE Transactions on Communications Technology*, 29(11), 644–654.
8. Rivest, R., Shamir, A., and Adleman, L. (1978). A method for obtaining digital signatures and public-key cryptosystems. *Communications for the ACM*, 21(2), 120–126.
9. McEliece, R. (1977). The theory of information and coding. In *Encyclopedia of mathematics and its applications*. Reading, MA: Addison-Wesley.
10. Koblitz, N. (1987). Elliptic curve cryptosystems. *Mathematics of Computation*, 48, 203–209.
11. Miller, V. (1986). Uses of elliptic curves in cryptography. *Advances in Cryptology—CRYPTO '85, Lecture Notes in Computer Science*, 218, 417–426.
12. Proos, J., and Zalka, C. (2003). Shor's discrete logarithm quantum algorithm for elliptic curves. *Quantum Information and Computation*, 3, 317–344.
13. Shor, P.W. (1997). Polynomial-time algorithms for prime factorization and discrete logarithms on a quantum computer. *SIAM Journal on Computing*, 26, 1484–1509.
14. Dodis, Y., Goldwasser, S., Kalai, Y., Peikert, C., and Vaikuntanathan, V. (2010). Public-key encryption schemes with auxiliary inputs. *Proceedings of Theory of Cryptography Conference*, 361–381.
15. Wu, W.W. (1985). *Elements of digital satellite communication*. Rockville, MD: Computer Science Press.
16. Kolawole, M.O. (2003). *Radar systems, peak detection and tracking*. Oxford: Elsevier Science.

2

<hr>

Satellites

<hr>

2.1 Overview

A satellite is a radiofrequency repeater. New-generation satellites are regenerative; that is, they have onboard processing capability making them more of an intelligent unit than a mere repeater (more is said of onboard processing in Section 2.9). This capability enables the satellite to condition, amplify, or reformat received uplink data and route the data to specified locations, or actually regenerate data onboard the spacecraft as opposed to simply acting as a relay station between two or more ground stations.

A typical satellite with onboard processors is the NASA Advanced Communications Technology Satellite (ACTS) shown in Figure 2.1. It was part of the payload on the Space Shuttle *Discovery* launched on September 12, 1993. According to NASA, its satellite weighs 3250 lb (1477.3 kg) and measures 47.1 ft (14.36 m) from tip to tip of the solar arrays and 29.9 ft (9.11 m) across the main receiving and transmitting antenna reflectors, with a height of 15.2 ft (4.63 m) from the spacecraft separation plane to the tip of the highest antenna. The solar arrays provide approximately 1.4 kW. The main communication antennas are a 7.2-ft (2.19-m) receiving antenna and a 10.8-ft (3.29-m) transmitting antenna. We describe more about the satellite components' design later in this chapter, particularly overall system design procedure, availability, and reliability in Section 2.6; antennas in Section 2.7; power systems in Section 2.8; onboard processing and switching systems in Section 2.9; and antenna control and tracking in Chapter 3, Section 3.4.2. Other characteristics of satellites are discussed in the next five sections.

As stated in Chapter 1, a satellite comprises several individual chains of equipment called a *transponder*, a term derived from *transmitter* and *responder*. The block diagram shown in Figure 2.2 may represent a transponder unit. As seen in the figure, a transponder may be described as a system composed basically of a bandpass filter required to select the particular channel's band frequencies, a frequency translator that changes frequencies from one level to another, and an output amplifier. Once amplified, the channels are recombined in an output multiplexer for the return transmission. All these devices must be stable over their operating temperature range to maintain

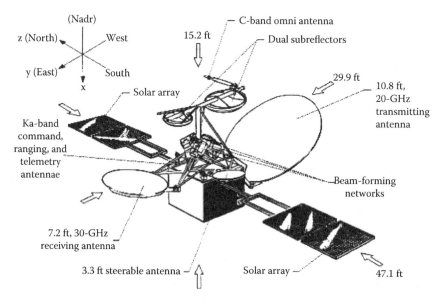

FIGURE 2.1
Geometry of a satellite. (Courtesy of NASA.)

the desired rejection characteristics. The functionality of these devices (each component block in Figure 2.2) is addressed later in this chapter. A transponder may channel the satellite capacity both in frequency and in power and may be accessed by one or by several carriers.

In most system applications, one satellite serves many earth stations. With the assistance of earth stations, fixed or transportable, satellites

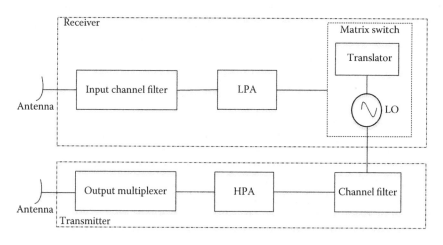

FIGURE 2.2
Basic transponder arrangement.

have opened a new era for global satellite multiaccess channels' data transmission and broadcast of major news events, live, from anywhere in the world. Commercial and operational needs dictate the design and complexity of satellites. The most common expected satellite attributes include the following:

1. Improved coverage areas and quality services, and frequency reusability
2. Compatibility of satellite system with other systems and expandability of current system that enhances future operations
3. High-gain, multiple hopping beam antenna systems that permit smaller-aperture earth stations
4. Increased capacity requirements that allow several G/s (gigabits per second) communication between users
5. Competitive pricing

Future trends in satellite antennas (concerning design and complexity) are likely to be dictated from the status of the satellite technology, traffic growth, emerging technology, and commercial activities.

The next two sections examine the type of satellites and the major characteristics that determine the satellite path relative to the earth. These characteristics are as follows:

1. Orbital eccentricity of the selected orbit
2. Period of the orbit
3. Elevation angle; the inclination of the orbital plane relative to the reference axis

2.1.1 Type of Satellites

There are, in general, four types of satellite:

Geostationary satellite (GEO)
High elliptical orbiting satellite (HEO)
Middle-earth orbiting satellite (MEO)
Low-earth orbiting satellite (LEO)

An HEO is a specialized orbit in which a satellite continuously swings very close to the earth, loops out into space, and then repeats its swing by the earth. It is an elliptical orbit approximately 18,000 to 35,000 km above the earth's *surface*, not necessarily above the equator. HEOs are designed to give better coverage to countries with higher northern or southern latitudes. Systems can be designed so that the apogee is arranged to provide continuous coverage in a particular area. By definition, an *apogee* is the highest-altitude point

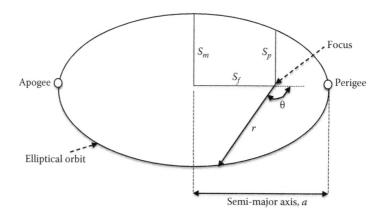

FIGURE 2.3
Geometric properties of an elliptical orbit (S_f = semifocal length, S_p = semiparameter, S_m = semi-minor axis, r = radius distance, focus to orbit path, θ = position angle).

of the orbit, that is, the point in the orbit where the satellite is farthest from the earth. To clarify some of the terminology, we provide Figure 2.3, which shows the geometric properties of an elliptical orbit. By geometry,

$$S_m = a\sqrt{1-e^2} \tag{2.1}$$

$$S_p = \frac{S_m^2}{a} = a\left(1-e^2\right) \tag{2.2}$$

where the eccentricity, or the amount by which the ellipse departs from a circle, is

$$e = \frac{S_f}{a} \tag{2.3}$$

The general equation of an ellipse can thus be written as

$$r = \frac{a\left(1-e^2\right)}{1+e\cos\theta} \tag{2.4}$$

It is apparent from (2.4) that if $e = 0$, the resulting locus is a circle.

An MEO is a circular orbit, orbiting approximately 8000 to 18,000 km above the earth's surface, again not necessarily above the equator. An MEO is a compromise between the lower orbits and the geosynchronous orbits. MEO system design involves more delays and higher power levels than satellites in the lower orbits. However, it requires fewer satellites to achieve the same coverage.

TABLE 2.1

Frequency Classification of LEOs

Type	Frequency	Usage
Big LEO	>10 GHz	Voice and data services. Requires more spectrums.
Little LEO	<10 GHz	Data services, in the lowest orbit. Often requires relatively small amounts of spectrum.

Source: Schulz, J.P., *CommLaw Conspectus*, 3, 185–186, 1995.

LEOs orbit the earth in grids that stretch approximately 160 to 1600 km above the earth's surface. These satellites are small, easy to launch, and lend themselves to mass production techniques. A network of LEOs typically has the capacity to carry vast amounts of facsimile, electronic mail, batch file, and broadcast data at great speed, and communicate to end users through terrestrial links on ground-based stations. With advances in technology, it will not be long until utility companies are able to access residential meter readings through an LEO system, or transport agencies and police are able to access vehicle plates, monitor traffic flow, and measure truck weights through an LEO system.

In the United States, the three satellites (HEO, MEO, and LEO) are collectively called LEOs, that is, low-earth orbiting satellite systems. By frequency designation, the LEOs are grouped as big and little LEOs as described in Table 2.1.

LEOs are subject to aerodynamic drag caused by resistance of the earth's atmosphere to the satellite passage. The exact value of the force caused by the drag depends on atmospheric density, the shape of the satellite, and the satellite's velocity. This force may be expressed in the form

$$F_d = -0.5\rho_a C_d A_{eq} v^2 \qquad \text{kg-m/s}^2 \qquad (2.5)$$

where ρ_a = atmospheric density (this density is altitude dependent, and its variation is exponential); C_d = coefficient of aerodynamic drag; A_{eq} = equivalent surface area of the satellite that is perpendicular to the velocity, v; and v = velocity of the satellite with respect to the atmosphere. The magnitude of this velocity is discussed in Section 2.2.

If the mass m_s of the satellite is known, the acceleration a_d due to aerodynamic drag can be expressed as

$$a_d = \frac{F_d}{m_s} \qquad \text{m/s} \qquad (2.6)$$

The effect of the drag is a decrease of the orbit's semimajor axis due to the decrease in its energy. A circular orbit remains as such, but its altitude decreases, whereas its velocity increases. Due to drag, the apogee in the

TABLE 2.2

Communication Satellite Frequency Bands Allocation

Band	Frequency Range (GHz)	Services
VHF	0.03–0.3	Messaging
UHF	0.3–1.0	Military, navigation mobile
L	1–2	Mobile, audio broadcast radiolocation
S	2–4	Mobile navigation
C	4–8	Fixed
X	8–12	Military
Ku	12–18	Fixed video broadcast
K	18–27	Fixed
Ka	27–40	Fixed, audio broadcast, intersatellite
Millimeter waves	>40	Intersatellite

elliptical orbit becomes lower, and as a consequence, the orbit gradually becomes circular. The longer the influence on the orbit, the slower the satellite becomes, and it eventually falls from orbit. Aerodynamic drag is more significant at low altitude (200 to 400 km) and negligible at about 3000 km because, in spite of the low value of atmospheric density encountered at the altitudes of satellites, their high orbital velocity, which is high, implies that perturbations due to drag are very significant.

A *geostationary* orbit is a nonretrograde circular orbit in the equatorial plane with zero eccentricity and zero inclination. The satellite remains fixed (stationary) in an apparent position relative to the earth, about 35,784 km away from the earth if its elevation angle is orthogonal (90°) to the equator. Its period of revolution is synchronized with that of the earth in inertial space. The geometric considerations for a geostationary satellite communication system will be discussed later in the text.

Commercial GEOs provide *fixed satellite service* (FSS) in the C- and Ku-bands of the radio spectrum. Some GEOs use the Ku-band to provide certain *mobile* services. The International Telecommunication Union (ITU) (see Chapter 7) has allocated satellite bands in various parts of the radio spectrum from very high frequency (VHF) to 275 GHz. Table 2.2 shows satellite communications frequency bands and the services they perform, while Table 2.3 shows typical link frequency bands.

TABLE 2.3

Typical Link Frequency Band Allocation

Band	Uplink Frequency (GHz)	Downlink Frequency (GHz)
C	5.925–7.075	3.7–4.2
Ku	14.0–14.5	11.7–12.2
Ka	27.5–31.0	17.7–21.2

Frequency bands in the ultra-high frequency (UHF) are suitable for communicating with small or mobile terminals, for television broadcasting, and for military fleet communication. The band of frequencies suitable for an earth-space-earth radio link is between 450 MHz and 20 GHz. Frequencies between 20 and 50 GHz can be used but would be subject to precipitation attenuation. However, if an availability greater than 99.5% is required, a special provision such as diversity reception and adaptive power control would need to be employed. Higher frequencies are more suitable for *intersatellite links* (ISLs) and may become usable as the orbital congestion arises at the lower frequencies. Another benefit of higher-frequency communication systems is that system components generally become smaller. For satellites, this translates to lighter weight, lower power, and reduced cost, and more importantly, it means increased mobility and flexibility.

2.2 Satellite Orbits and Orbital Errors

For geometric consideration, a satellite can also be explained as a body that moves around another body (of greater mass) under the influence of the gravitational force between them. The force F required to keep a satellite in a circular orbit can be expressed as

$$F = m_s \omega^2 r \qquad \text{N (Newton)} \qquad (2.7)$$

where m_s = mass of the satellite, g; ω = angular velocity of the satellite, rad/s; and r = radius of the orbit, m, = $R_e + h_0$, m.

This is the distance of a synchronous satellite from the center of the earth. h_0 is the orbit altitude, that is, the height above the subsatellite point on the earth terminal (m).

The gravitational force F_g acting on the satellite of mass m_s at distance r from the center of the earth is

$$F_g = m_s g \frac{R_e^2}{r^2} \qquad \text{N} \qquad (2.8)$$

where g = acceleration due to gravity at the surface of the earth = 9.807 m/s², and R_e = radius of the earth. The value varies with location. For example,

$$R_e \text{ at the equator} = 6378.39 \text{ km } (\approx 6378 \text{ km})$$

$$R_e \text{ at the pole} = 6356.91 \text{ km } (\approx 6357 \text{ km})$$

Consequently, for a satellite in a stable circular orbit round the earth,

$$F = F_g \tag{2.9}$$

In view of (2.7) and (2.8) in (2.9),

$$r^3 = g \frac{R_e^2}{\omega^2} \tag{2.10}$$

The period of the orbit, t_s, that is, the time taken for one complete revolution (360° or 2π radians), can be expressed as

$$t_s = \frac{2\pi}{\omega} = \frac{2\pi}{R_e} \sqrt{\frac{r^3}{g}} \qquad \text{s} \tag{2.11}$$

If we assume a spherical homogeneous earth, a satellite will have an orbital velocity represented by

$$v = R_e \sqrt{\frac{g}{r}} \qquad \text{m/s} \tag{2.12}$$

For elliptical orbits, Equations (2.11) and (2.12) are also valid by equating the ellipse semimajor axis a with the orbit radius, r (i.e., $r = a$). In terms of the orbit parameters, r is replaced with the average of the apogee to the focus and the perigee to the focus. By definition, a *perigee* is the lowest-altitude point of the orbit, whereas an *apogee* is the highest-altitude point of the orbit. In a circular orbit, with variable altitude and upon substitution of empirical values in (2.11) and (2.12), Figure 2.4, which relates period

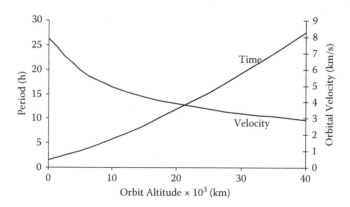

FIGURE 2.4
Satellite period and orbital speed vs. altitude.

and velocity for circular orbits, assuming a spherical homogeneous earth, is plotted.

For a circular orbit at an altitude of 35,784 km, Figure 2.4 shows that a geosatellite orbit takes a period of rotation of the earth relative to the fixed star (called *sidereal day*) in 86,163.9001 s, or 23 h 56 min 4 s. In some books and papers, an approximate value of 36,000 km is frequently cited for the altitude of the satellite in geosynchronous orbit. The geosynchronous orbit in the equatorial plane is called the *geostationary orbit*. Although a satellite in the geostationary orbit does not appear to be moving when seen from the earth, its velocity, from Figure 2.4, in *space* equals 3.076 km/s (11,071.9 km/h).

Low-altitude satellites, which have orbits of less than nominally 24 h, have other applications in addition to those earlier tabulated in Table 2.1. The applications include reconnaissance purposes, provision for communications at extreme north and south latitudes when in a polar orbit, and numerous business opportunities in producing remotely located monitoring and data acquisition devices that could be accessed by satellite.

2.2.1 Orbital Errors

It is not possible to put a satellite into a perfect geostationary orbit because any practical orbit is slightly inclined to the equatorial plane. In addition, it is not exactly circular; it does not have exactly the same period as that of the earth's rotation, and it is constantly bombarded by disturbing forces (such as the attraction of the sun and moon) that try to change the orbit. These disturbing forces cause satellites to drift slowly in longitude. Their effects are counteracted from time to time by operating thrusters on the satellite. It is logical to suggest that these forces introduce orbital errors outside the intended nominal longitudes.

Advances in space technology have enabled minimization of the orbital errors. For example, INTELSAT V satellites were kept within ±0.1° of the equator and of their nominal longitudes, whereas INTELSAT VI satellites are kept within ±0.02° of the equator and ±0.06° of their nominal longitudes.

2.3 Coverage Area and Satellite Networks

2.3.1 Geometric Coverage Area

Clarke [2] foresaw in his article that it would be possible to provide complete radio coverage of the world from just three satellites, provided they could be precisely placed in geosynchronous orbit. Figure 2.5 demonstrates this.

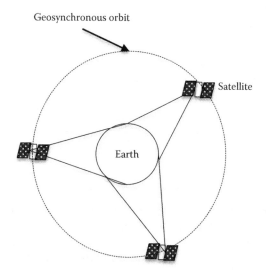

FIGURE 2.5
Complete coverage of the earth's surface from three satellites.

The amount of coverage is an important feature in the design of earth observation satellites. Coverage depends on altitude and look angles of the equipment, among several factors. To establish the geometric relationship of the coverage, we take a section of the satellites in Figure 2.5 as an illustration. This section is redrawn as shown in Figure 2.6.

The maximum geometric coverage can then be defined as the portion of the earth within a cone of the satellite at its apex, which is tangential to the earth's surface. Consider the angle of view from the satellite to the earth terminal as α; then the apex angle is 2α. The view angle has a mathematical physical function given by

$$\alpha = \sin^{-1}\left(\frac{R_e}{h_o + R_e}\right) = \sin^{-1}\left(\frac{R_e}{r}\right) \tag{2.13}$$

Using empirical values ($R_e = 6378$ km, $r = 42,162$ km), the apex angle 2α equals $17.33°$, the planar angle beamwidth. It follows that an "earth coverage" satellite antenna must have a minimum beamwidth θ_{BW} of $17.33°$. In practice, an antenna of $18°$ or $19°$ beamwidth is used to allow for directional misalignment. Thus, for a single geostationary satellite to illuminate in excess of a third of the earth's surface, the antenna minimum beamwidth must be at least 2α.

The beamwidth of the satellite antennas determines the area of the earth serviced or covered. The beamwidth required directly determines the antenna gain and, for a given operating frequency, the physical size of the antenna aperture (see a further discussion on antennas in Section 2.7).

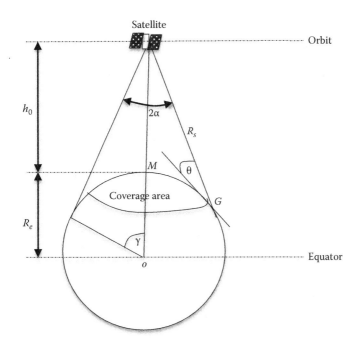

FIGURE 2.6
An illustration of coverage area and apex angle.

Using the notations in Figure 2.6 as a guide, the coverage area, A_{cov}, from which the satellite is visible with an elevation angle of at least θ can be established as

$$A_{cov} = 2\pi R_e^2 (1 - \cos \gamma) \tag{2.14}$$

where γ is the central angle. It is a spherical trigonometric relation that relates to the earth and satellite coordinates (to be discussed in detail in Section 2.4).

The apex angle required at the satellite to produce a given coverage A_{cov} must satisfy

$$2\pi\{1 - \cos \alpha\} = \frac{A_{cov}}{h_0^2} \tag{2.15}$$

However, for small angles, that is, $\alpha \ll 1$, we can approximate the global beamwidth to

$$2\alpha \approx \frac{d_{cov}}{h_0} \tag{2.16}$$

where d_{cov} is the coverage area diameter.

In order to provide communications among areas serviced by the three satellites, a terrestrial repeater must be provided at a location where both satellites are visible. Alternatively, a link can be established between the satellites (see Section 2.3.2) by using multibeam satellite configuration (see Section 2.3.3). Also, by using nonsynchronous orbits, the trade-off between synchronous and nonsynchronous orbits is a complicated system design problem.

2.3.2 Satellite Constellation

A *constellation* is a group of similar satellites working together in partnership to provide a network of useful service. The constellation (or configuration) of satellites in the LEO system is designed to function as a network primarily to get more or better coverage. Each satellite in an orbital plane maintains its position in relation to the other satellites in the plane. Each satellite in an LEO constellation, for example, acts as a switching node, and each satellite is connected to nearby satellites by *intersatellite links*. An example is the two LEOs cross-linked in Figure 2.7. These links route information received from a user terminal or a network of user terminals through the satellite network

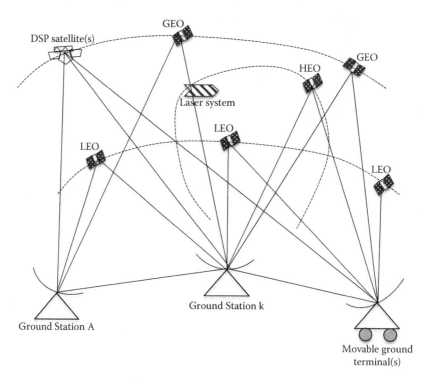

FIGURE 2.7
A sketch of SBIRS: an example of intersatellite links and satellite constellation.

and eventually toward a gateway to an earth-based network or a mobile user. Each satellite node implements signaling and control functions that set up and release connections. It is worth noting that the signaling protocol used when satellite systems are constellated is similar to the standard public switched telephone network/integrated services digital network (PSTN/ISDN) and SS7 protocol for cellular mobile systems (Chapter 8 describes more about the protocols and cellular mobile system).

Satellite networks often require communications between two satellites (like the LEOs in Figure 2.7) via an intersatellite link (ISL). As a communication link, an ISL has the disadvantage that both transmitter and receiver are spaceborne, thereby limiting operation to both low transmitter power and low figure-of-merit values. What is the trade-off? The answer will become obvious in Section 4.2.1 of Chapter 4.

Typical examples of satellite constellations are the U.S. Defense satellite operational networks called FLTSATCOM (fleet satellite communication system) and SBIRS (space-based infrared system) and the Iridium and Teledesic networks, which provide global mobile telephone and paging services.

FLTSATCOM provides worldwide communications for U.S. Navy and Air Force units, except for the polar regions. FLTSATCOM comprises four satellites. Each satellite is injected into a near-geosynchronous equatorial orbit and positioned at longitudes 100°W, 23°W, 71.5°E, and 172°E. Each satellite overlaps in coverage with the adjacent satellite, thereby avoiding any gap in space segment continuity. The coverage is between the latitudes of 70°N and 70°S. The fleet constitutes the primary Navy's broadcast and ship interchange communication system. It also provides vital communications to the Allied Forces worldwide.

SBIRS is proposed to be the next frontier of the U.S. missile defense system: a sort of missile shield [3]. SBIRS will comprise a network of satellites: LEO, HEO, and GEO (see Figure 2.7). In theory, the GEO forms the frontline satellites that provide the first warning of missile launches over the equator, and HEOs cover the north pole. The information received by the frontline satellites is then passed via dedicated defense support program (DSP) satellites to earth terminals. The DSP satellites are programmed to look for the launch flares of a missile taking off or the distinctive double flares that mark the explosion of a nuclear weapon. The details of the missile trajectories are then passed to a network of LEOs, which would track the missile warheads after they separate from their launchers and cruise through space. The space-based laser system is then activated to track incoming missiles.

Lasers can be used to communicate between different satellites in space, and between terrestrial devices and satellites. Using a laser terminal to communicate with a satellite provides higher speeds, larger bandwidth, and greater security than communicating using radiowaves, but it carries several restrictions as well. For instance, lasers require a direct line of sight to communicate with any satellite. To the contrary, radiowaves can penetrate many barriers that are impossible for lasers to penetrate, including in adverse weather conditions.

The Iridium network, deployed in 1998, consists of 66 satellites in six orbital planes [4]. Each satellite has 48 spot beams. The satellite network's switching algorithm allows some of the beams to be switched off as it approaches the poles, but maintains a sizable proportion of the beams at any given time. Each satellite is connected to four neighboring satellites, and the intersatellite links operate in the 23.18 to 23.38 GHz band. Communications between the satellite network and terrestrial network are through ground station gateways. The uplink and downlink frequency bands to the gateways are in the 29.1 to 29.3 GHz and 19.2 to 19.6 GHz ranges, respectively.

The Teledesic network plans for 288 satellites to be deployed in a number of polar orbits [5]. Each satellite is interconnected to eight adjacent satellites to provide tolerance to faults and adaptability to congestion. In this network the earth is divided into approximately 20,000 square supercells, each 160 km long and comprised of 9 square cells. Each satellite's beam covers up to 64 supercells; the actual effective supercell coverage depends on the satellite's orbital position and its relative distance to other satellites. Each cell in the supercell is allocated a time slot, and each satellite focuses on the cell in the supercell at that allotted time. When the beam is directed at a cell, each terminal in the cell transmits on the uplink, using one or more frequency channels that have been assigned to it. During this time frame, the satellite transmits a sequence of packets of terminals in the cells. The terminals in turn receive all packets and distribute to respective localities.

2.3.3 Multibeam Satellite Network

Multibeam antennas carried aboard the satellites attempt to conserve available frequencies. A multibeam antenna transmits a family of pencil-thin beams, often so small that by the time they reach earth's surface, their footprint covers an oval only a few tens of kilometers wide. As an illustration, instead of a global angular width of 17.33° for one satellite for a global beam coverage discussed in Section 2.3.1, multiple narrow beams, each of an angular width of 1.73° with reduced coverage and increased gain, are used. This scheme permits multibeam satellite configuration. This scheme includes the following advantages:

1. Power is divided among the beams, and the bandwidth remains constant for each beam. As a result, the total bandwidth increases by the number of beams.

2. Performance improves as the number of beams increases, although limited by technology and the complexity of the satellite, which increases with the number of beams.

3. There is extended satellite coverage from the juxtaposition of several beams, and each beam provides an antenna gain that increases as the angular beamwidth decreases.

4. Frequency reuse is achieved, which means using the same frequency band several times so as to increase the overall capacity of the network without increasing the allocated bandwidth. This can be achieved by exploiting the isolation resulting from antenna directivity to reuse the same frequency band in different beams. For instance, an antenna designed to transmit or receive an electromagnetic wave of a given polarization can neither transmit nor receive in the orthogonal polarization. This property enables two simultaneous links to be established at the same frequency between two identical locations. This process is called *frequency reuse* or *orthogonal polarization*. To achieve this feat, either two polarized antennas must be provided at each end or, preferably, one antenna, which operates with the two specified polarizations, may be used. The drawback is that this could lead to mutual interference of the two links. The magnitude of this interference on the total transmission link will be examined in Chapter 4.

2.4 Geometric Distances

By considering the geometry of the geosatellite's orbit in its orbital plane, we will be able to calculate:

1. The distance between the satellite and earth station, called *slant range*, R_s.

2. The azimuth and elevation angles, collectively called the *look angles*. The look angles are the coordinates to which an earth station antenna must be pointed to communicate with a satellite. *Azimuth angle* a_z is the angle at which the earth station's disk is pointing at the horizon, whereas the *elevation angle* θ is the angle by which the antenna boresight must be rotated to lock on to the satellite.

3. The width of the viewed section along the orbit ground trace, called *swath distance* or *swath width*.

For clarity and without the coverage area, Figure 2.6 is redrawn as Figure 2.8. Using Figure 2.8 as a guide, we can establish the expressions governing most of the above-listed parameters.

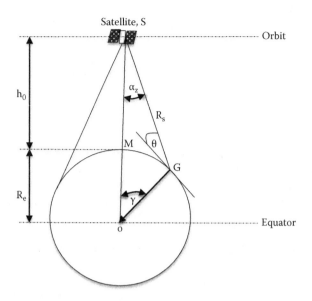

FIGURE 2.8
Geometry of look angles.

In the figure, S = position of satellite, G = position of earth station, $R_v = OG$ = geocentric radius of earth at G latitude, θ = elevation angle of satellite from the earth station, L_{ET} = latitude of the earth station (this value is positive for latitudes in the northern hemisphere [i.e., north of the equator] and negative for the southern hemisphere [i.e., south of the equator]), M = location of sub-satellite point (this location's longitude and latitude are determined from a satellite ephemeris table; nominally, latitude is taken as 0° for a geostationary satellite), L_{SAT} = latitude of the satellite, Δ = difference in longitude between the earth station and the satellite, γ = central angle, and r = radius of the orbit = $OM + MS = R_e + h_0$.

Using the spherical trigonometric relations, we determined the central angle from

$$\gamma = \cos^{-1}\left(\sin L_{SAT} \sin L_{ET} + \cos L_{SAT} \cos L_{ET} \cos \Delta\right) \tag{2.17}$$

Using the cosine law, we find the slant range equation to be

$$R_s = \sqrt{R_e^2 + r^2 - 2rR_e \cos \gamma} \qquad \text{km} \tag{2.18a}$$

Alternatively,

$$R_s = \frac{R_v \sin \gamma}{\cos(\gamma + \theta)} \qquad \text{km} \tag{2.18b}$$

The geocentric radius R_v may be described by

$$R_v = R_e\left(0.99832 + 0.001684\cos 2L_{ET} - 0.000004\cos 4L_{ET}\cdots\right) \qquad \text{km} \quad (2.19)$$

The elevation angle θ may be written as

$$\theta = \tan^{-1}\left(\frac{\cos\Delta\cos L_{ET} - \left(R_e/r\right)}{\sqrt{1 - \cos^2\Delta\cos^2 L_{ET}}}\right) \qquad \text{deg} \qquad (2.20a)$$

Alternatively,

$$\theta = \tan^{-1}\left(\cot\gamma - \frac{R_v}{r\sin\gamma}\right) \qquad \text{deg} \qquad (2.20b)$$

The above equations developed for elevation angle θ give the geometric value. The true elevation angle θ_t, taking into account the average atmospheric refraction, is given by the approximate equation [11]:

$$\theta_t = 0.5\left(\theta + \sqrt{\theta^2 + 4.132}\right) \qquad \text{deg} \qquad (2.21)$$

and the azimuth angle is

$$a_z = 180 + \tan^{-1}\left(\frac{\tan\Delta}{\sin L_{ET}}\right) \qquad \text{deg} \qquad (2.22a)$$

Alternatively,

$$a_z = 180 + \frac{-\sin\Delta}{\sqrt{1 - \cos^2 L_{ET}\cos^2\Delta}} \qquad \text{deg} \qquad (2.22b)$$

For the southern hemisphere azimuth angle, the 180 term is deleted. The magnetic heading of the antenna can be expressed as

$$\alpha_H = a_z + \Delta_\alpha \qquad \text{deg} \qquad (2.23)$$

where Δ_α is the deviation, as the angle between the true north pole and the north magnetic pole.

Example 2.1

An earth station antenna is located at 89°W longitude and 35°N latitude. It is intended to transmit and receive signals to a satellite in geostationary orbit at 82°W longitude. Assume a deviation angle of –3.3°. Calculate the compass heading to which the antenna would be pointing and its elevation.

SOLUTION

$$\Delta = 82 - 89 = -7°$$

$$L_{ET} = 35°$$

From (2.20a), the elevation angle

$$\theta = \tan^{-1}\left(\frac{\cos(-7)\cos(35) - 0.1512}{\sqrt{1 - \cos^2(-7)\cos^2(35)}} \right) = 48.67°$$

Using (2.21), the true elevation is calculated as 48.67°.
    The azimuth angle is calculated using (2.21b):

$$a_z = 180 + \frac{-\sin\Delta}{\sqrt{1 - \cos^2 L_{ET} \cos^2 \Delta}} = 168°$$

Consequently, the compass heading is estimated using (2.23):

$$\alpha_H = a_z + \Delta_\alpha = 164.7°$$

We can conclude that the antenna must be pointed toward a compass heading of 164.7° and be adjusted to an elevation of 48.67° to transmit or receive signals from the satellite in the geostationary orbit.

2.5 Swath Width, Communication Time, and Satellite Visibility

2.5.1 Visibility of Satellites from the Earth

One of the most important pieces of information about a communication satellite is whether it can be seen from a particular location on the earth. The use of (2.21) and (2.22b) enables us to interpret the bearings relative to the local meridian. As an illustration, suppose $L_{ET} = 40°$ and $\Delta = 50°$. Consequently, $a_z = 62°$ and the elevation of the satellite $\theta = 21°$. Now, on the one hand, suppose the satellite is east of the earth station and the station is south of the equator; the azimuth of the satellite from the station will equal a_z, that is, 62°E of N. On the other hand, if the station is north of the equator, then the azimuth of the satellite is $(180 - a_z) = 118°$E of N. Similarly, if the satellite is west of the station, then the azimuth of the satellite is $(180 + a_z) = 242°E$ of N for the station north of the equator, and finally, $(360 - a_z) = 298°E$ of N for a station south of the equator.

2.5.2 Communication Time

The amount of time t_p an earth station is required to communicate with an orbiting satellite as it passes overhead is given by

$$t_p = \left(\frac{\gamma}{180°}\right)\left(\frac{t_s}{1 \mp \left(t_s \middle/ t_e\right)}\right) \tag{2.24}$$

where t_s = orbit period and t_e = rotation period of the earth = 1 sidereal day. Note that the sign $\mp$ depends on whether the satellite is in a retrograde (opposite-direction) orbit or a prograde (same-direction) orbit. For instance, if a satellite is in a prograde orbit, $t_e = t_s$, and as such $t_p = \infty$. The expression (2.24) is valid for a satellite orbiting at altitude h_0 and passing over any point on earth with an elevation angle exceeding θ.

2.5.3 Swath Width

The width of the viewed section along the orbit ground trace is called the *swath width*. Following the previous assumption of a spherical earth, it can be demonstrated that the swath width may be written as

$$w_s = \frac{R_e}{90}\left(90 - \frac{\theta_{BW}}{2} - \cos^{-1}\left(\frac{h_o R_v}{R_e}\sin\frac{\theta_{BW}}{2}\right)\right) \tag{2.25}$$

where θ_{BW} is the antenna beamwidth in degrees. Other symbols are as defined previously. The above expression can be used to determine the coverage at any latitude during a particular operation. For practical explanation, when a polar orbit is used, the percentage coverage at the equator is taken as the ratio of swath width to the total distance between consecutive passes. To the contrary, at latitudes beside the equator, the percentage coverage in a given day is the equatorial coverage divided by the cosine of the latitude.

2.6 Systems Engineering: Design Procedure

The constraints imposed by the satellite on size, shape, cost, and weight are important factors in design consideration [6, 12]. A satellite design block diagram is shown in Figure 2.9; it shows the flow-down of requirements to

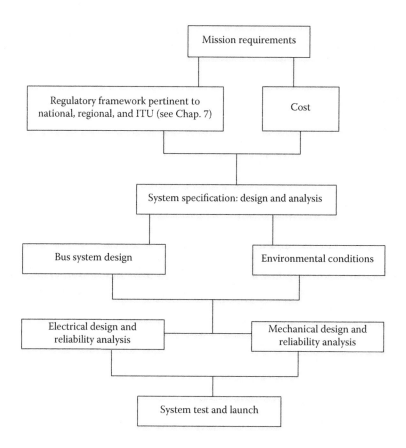

FIGURE 2.9
Design flowchart of a satellite system.

system development. Several factors dictate the satellite system design process, but some generalizations are possible.

The overriding concern is the cost for developing the system. It is imperative that system cost-effectiveness analysis be performed, as it supports the development of the life cycle of products, processes, and risk management activities.

2.6.1 Mission Requirements

In general, mission requirements are intended to indicate, as definitively as practical, the objectives of the mission: what is to be achieved and how to measure performance. In some cases, the requirements are not clear-cut. It is the system engineer's responsibility to (1) ensure that both the customer and the developer understand the mission objectives and (2) clearly define practically attainable and practically verifiable results.

2.6.2 Regulatory Authorities

Regulatory procedures have been adopted by each country in conformity with ITU provisions (to which member states agree). These procedures are intended to ensure that radiocommunication services operating at any given time do not cause or receive harmful interference. Each country's regulatory agency must make specific regulations concerning domestic satellite networks and earth terminals. The designer must comply with these during the system design phase, and the operator system must follow them during the implementation phase. For example, the Australian Communications Authority (ACA) and the Federal Communications Commission (FCC) are radiocommunications regulators of Australia and the United States, respectively. Chapter 7 defines the relationship between international regulators (ITU) and national or regional regulators.

2.6.3 System Specifications: Analysis and Design

System specifications are essentially sets of dos and don'ts as applied to system operations and could be labeled as technical requirements in terms of function, performance, interface, and design requirements for the system, equipment, and software. The technical requirements may relate to manning, operating, maintaining, and logistically supporting the system, equipment, or software to the extent that these requirements define or constrain the design.

It is at this phase that the requirements are dissected and analyzed in greater detail, covering:

1. Description of the methods, procedures, and tools for analysis of mission statement(s), operations, and environment
2. Identification of functional and performance requirements for design, development, manufacturing, verification, deployment, operations, support, training, and disposal
3. Determination of system constraints, including descriptions of the following:

   Reliability and maintainability

   Survivability

   Electromagnetic compatibility, radiofrequency management, and electrostatic discharge

   Human engineering and human systems integration

   Safety and health hazards

   Test and evaluation

   Integrated diagnostics

   Other system functionality that has bearing on the determination of performance and functional requirements for the system

4. Risk management, covering at least the following:

Approach and criteria for risk identification, prioritization, sensitivity assessment, handling, and risk impact integration into the decision process

The risks associated with the system development test and evaluation requirements

Appropriate plans to handle technical risks (for example, those associated with technology and integration verification, backup development, etc.)

Risk control and monitoring measures, including technical performance measurement

Risks associated with cost and schedule performance measurement

5. Environmental factors

The prime environmental problems a satellite encounters are electromagnetic charges, radiation, pressure differential, solar radiation or flare, and other artifacts, like meteoroids. The space environment is composed of electromagnetic and corpuscular radiation. Electromagnetic radiation comes mainly from the sun and exhibits the properties of waves traveling at the speed of light. Corpuscular radiation originates both in the sun and in galactic space and consists of charged and uncharged particles whose velocity is the product of the mechanics of generation. In essence, environmental conditions during launch and in space affect mechanical and electrical design as well as reliability consideration.

Upon completion of these analyses, an *operational concept document* (OCD) is produced, whose main purposes are to:

1. Specify the system characteristics from an operational perspective

2. Ensure that users understand the system goals

3. Provide guidance for development of subsequent system definition documents, such as system specification, interface specification, product specification, test specification, etc.

4. Describe the user organization and mission from an integrated system or user point of view

It is important to note that, at this system-level design, the customer's approval must be gained for the OCD to become valid because the customer is the owner of the OCD. The production and subsequent approval of the OCD is the first step in clarifying the customer's needs and establishing the customer's requirements as well as providing practical guidelines on how to apply these outlined requirements for system analysis and design.

2.6.4 System Design: Bus, Electrical, and Mechanical

The satellite bus system design must consider the beam pointing accuracy, velocity, and angular range.

Like all materials that require coupling, all mechanical conditions must be properly observed. These conditions include accurate dimensioning, overall weight of the component parts, sinusoidal and random vibration, noise, heat, dynamic strength, stiffness, acceleration, and launching (or deployment) shock. These conditions, in some regards, apply to the electrical component's design.

To reduce distortions due to heat, the material property must be well determined and stress release design properly considered. Modern design incorporates venting to avoid destruction due to pressure change during satellite launch.

The electrical design must first consider what type of antenna is preferred: omnidirectional, single beam, multibeam, or shaped beam. *Omnidirectional* antennas are those that radiate uniformly in all directions. *Single-beam* antennas are used as global beam or spot beam antennas in satellite communication antennas. Horn antennas are generally used for global beam purposes. Spot beam antennas usually use larger-aperture diameters to form a narrow beam; these antennas usually employ reflectors (e.g., paraboloidal reflector, horn reflector, Cassegrain reflector, and Gregorian reflector). The characteristics of these antennas, which are generally applicable to both satellite and earth stations, are discussed in Section 2.7. *Shaped-beam* antennas are antennas whose radiated beams are shaped to a prescribed pattern. A reflector or lens antenna is widely used with multiple-feed elements for generating a shaped beam for a satellite communication antenna system. The advantage of beam shaping includes enhancement of the antenna gain and improvement of beam-to-beam isolation, which is a key parameter for increasing communication capacity due to frequency reuse.

As demonstrated in Section 2.3, the beamwidth of the satellite antenna is directly related to the size of the coverage area and, for a given operating frequency, the antenna aperture diameter D. Antenna selection is governed by the aperture-frequency (D^*f) product, as shown in Table 2.4. The operating frequency f is in GHz and the aperture D is in m.

After selecting the appropriate antenna, then, for a fixed received carrier level and the noise (C/N) and prechosen modulation system, calculate the bandwidth, number of beams, and frequency or frequency band necessary for a coverage area of interest. Next, determine the desired beam shape and polarization and their isolation. A typical isolation requirement is about 27 dB for circular polarization and 33 dB for linear polarization. It should be noted that low sidelobes and cross-polarization are necessary to prevent excessive interference among beams. The beamwidth or gain and the frequency determine the initial value of the aperture diameter D. At this stage,

TABLE 2.4

Antenna Selection Criteria

D^*f	Type of Antenna
0.03–0.3	Omnidirectional
0.15–3.0	Horn
3.0–180.0	Paraboloidal reflector
0.9–180.0	Offset—paraboloidal reflector
9.0–330.0	Cassegrain or Gregorian reflector
0.9–330.0	Offset—Cassegrain or Gregorian reflector
0.9–360.0	Array (phased or planar)

decide on feed type (e.g., horn, waveguides, etc.). Do some prototyping of the design and perform a parametric study to estimate the optimum antenna parameters. Then verify the validity of your design.

Generally speaking, electrical design of satellite antennas must take into account the launch and space environments. The launch environment is often neglected because the antenna is disabled during this phase. Despite this, certain precautions must be taken. As noted by Hwang [7], for instance, the antenna must be designed against:

1. Gas discharge at pressure ($\approx$0.5 mmHg) for 60 km or lower altitude because the charge induced by charged particles is released as electrostatic discharge. The discharge often leads to electrostatic breakdown of antenna materials and degradation of communication quality.
2. Multipactor breakdown at lower pressure ($\leq$0.001 mmHg) for 90 km or higher altitude; caused by secondary emission of electrons in microwave circuits.
3. Also, a satellite antenna must be designed to withstand the dynamic thermal and mechanical stresses of the satellite environment.

2.6.5 System Reliability and Availability Analyses

System availability can be defined as the availability of the link from the transmitting earth station up through the satellite and down to the receiving earth station. System availability is also defined as the proportion of time in some long interval the system is working. The two definitions demand two different analyses.

By the first definition, the system availability expression can be established by assuming that the system is made up of independently operating subsystems—the transmitting earth station, receiving earth station, and satellite. As such, a link outage, an earth station failure, and a satellite failure occur mutually independently. The system engineer must first determine the link availability, P_{LA}, which is the average probability of the bit error exceeding design threshold, P_{th}. This threshold is the sum of the uplink and downlink threshold outages, neglecting joint threshold outages. If we let P_0 represent the total percentage of outages per year above the threshold, then the system link availability can be expressed as

$$P_{LA} = 1 - P_0 \qquad (2.26)$$

Thus, the system availability

$$P_{SA} = P_{ESA} P_{LA} P_{SAT} P_{ESB} \qquad (2.27)$$

where P_{SA} = system availability, P_{ESA} = availability of transmitting earth station A, P_{ESB} = availability of receiving earth station B, and P_{SAT} = satellite availability.

By the second definition, system availability expression can be established by also assuming that the system is made up of independently operating subsystems—the transmitting earth station, receiving earth station, and satellite. These subsystems fail independently and with adequate facilities to initiate repair of any of the subsystems when they fail. The failure and repair process of each subsystem is such that with each subsystem can be associated an uptime, a downtime, and a time between failures, which all have a constant mean value when the system has been operating long enough to be in statistical equilibrium. Then the means of these times can be denoted by MUT, MDT, and MTBF, respectively. This analysis assumes that:

1. The system-associated times depend on external events (e.g., raining, not raining, daytime, nighttime, eclipse, solar flare).
2. The means of these times can be defined for these events.
3. Online replacement of failed units by functioning units is in place.

Then, the system availability can be expressed as

$$P_{SA} = \frac{MUT}{MTBF}$$

$$= \frac{MTBF - MDT}{MTBF} \qquad (2.28)$$

FIGURE 2.10
System reliability block diagram.

Similarly, the system unavailability can be written as

$$\overline{P}_{SA} = 1 - P_{SA}$$

$$= \frac{MDT}{MTBF}$$

(2.29)

The system reliability block diagram can be drawn, as in Figure 2.10, for satellite communication between two earth stations (A and B) to be established.

The system availability can be expressed as a product of each subsystem's availability:

$$P_{SA} = \left(\frac{MUT_{ESA}}{MTBF_{ESA}} \right) \left(\frac{MUT_{SAT}}{MTBF_{SAT}} \right) \left(\frac{MUT_{ESB}}{MTBF_{ESB}} \right)$$

(2.30)

Subscripts *ESA*, *SAT*, and *ESB* denote earth station A, satellite, and earth station B, respectively. If r out of n units $X_1, X_2, X_3, ..., X_n$ are required for each subsystem to be working, and each unit X_i has the same availability P_{ai}, then the subsystem availability can be written as

$$P_{ASub} = \binom{n}{r} P_{ai}^r \overline{P}_{ai}^{n-r} + \binom{n}{r+1} P_{ai}^{r+1} \overline{P}_{ai}^{n-r-1} + \cdots$$

$$+ \binom{n}{r+j} P_{ai}^{r+j} \overline{P}_{ai}^{n-r-j} + \cdots + P_{ai}^n$$

(2.31)

where (n/m) are the binomial coefficients.

Example 2.2

The mean times per year of earth stations A and B and the satellite are collated as tabulated below. Calculate the system availability.

Subsystems	MTBF (h/year)	MDT (h/year)
Earth station A	300	3
Earth station B	200	10
Satellite	500	4

SOLUTION

From (2.28), the availability of each subsystem can be estimated:

$$\text{Earth station A:} \quad P_{aA} = \frac{300 - 3}{300} = 0.99$$

$$\text{Earth station B:} \quad P_{aB} = \frac{200 - 10}{200} = 0.95$$

$$\text{Satellite:} \quad P_{aSAT} = \frac{500 - 4}{500} = 0.992$$

Hence, the system availability from (2.30) is

$$PSA = P_{aA}P_{aB}P_{aSAT} = 0.933$$

This means that the system will be available for 93.3% (alternatively, unavailable for about 6.7%) of the year.

2.6.5.1 System Reliability

System reliability attempts to predict the future or quantitatively provides a certain level of guarantees the system will meet, a by-product of the reliability theory. By definition, the reliability $R(t)$ of a subsystem is the ratio of the number of surviving components $N_s(t)$ within a system at time t to the number of components at the start of the subsystem test period N_0; that is,

$$R(t) = \frac{N_s(t)}{N_0} \tag{2.32}$$

The number of components that failed, $N_f(t)$, at time t can be written as

$$N_f(t) = N_0 - N_s(t) \tag{2.33}$$

From a system engineering perspective, we can relate any component failure to the system MTBF. To do this, we need to know the probability of any one of the components failing at the start. If one assumes that all components were continuously tested until all of them failed, the MTBF of the ith component failing after time t_i can be written as

$$MTBF = \frac{1}{N_0} \sum_{i=1}^{N_0} t_i \tag{2.34}$$

In fact, MTBF is the reciprocal of the average failure rate λ_r. For simplicity, the average failure rate is assumed constant for a given batch of components. Thus, for a small sample of failed components, we can write

$$\lambda_r = \frac{1}{N_s} \frac{\Delta N_f}{\Delta t} = \frac{1}{N_s} \frac{dN_f}{dt} \tag{2.35}$$

Alternatively, λ_r can be expressed in terms of rate of survival (dN_s/dt) as follows. Differentiating (2.33) with respect to t, we found that dN_s/dt is the negative of dN_f/dt. By substituting the differentiation in (2.35) and in view of the reliability definition of (2.32), the average failure rate is

$$\lambda_r = -\frac{1}{N_s}\frac{dN_s}{dt}$$

$$= -\frac{1}{R(t)}\frac{dR(t)}{dt} \tag{2.36}$$

A solution for reliability $R(t)$ in (2.36) may be written as

$$R(t) = e^{-\lambda_r t} \tag{2.37}$$

This expression shows that reliability decreases exponentially with time, with zero reliability after infinite time ($t \geq \infty$). In practice, end of useful life is often taken to be the time t at which reliability falls to 0.3679 (i.e., $1/e$), which is when $t = $ MTBF. An example of a reliability problem is given in Section 2.8 (Example 2.6).

2.6.5.2 Redundancy

Redundancy is provided whenever failure of a unit or subsystem will cause the loss of a significant part of the satellite communication capacity. A transponder shown in Figure 2.2, for example, will have two parallel pairs of low-noise amplifier (LNA) in the receive chain and two parallel pairs of high-power amplifier (HPA) in the transmit chain, as shown in Figure 2.11. If one LNA is lost, the transmitted signals from the antenna that pass through the LNA would be recombined at the output of the other parallel LNA to avoid catastrophe. Similarly, if one HPA is defective, the other HPA will allow information to be transmitted.

2.6.6 System Integration, Test, and Evaluation

A well-thought-out plan should be in place that details how the integration and assembly of the system will be undertaken, with emphasis on risk management and continuing verification of all external and internal interfaces (physical, functional, and logical). The plan should allow a demonstration of upward and downward traceability of requirements with system engineering process inputs.

The system test should encompass detailed functional and performance measurements and evaluation of all quantifiable characteristics at all system levels. To test the system, essential system operational attributes through

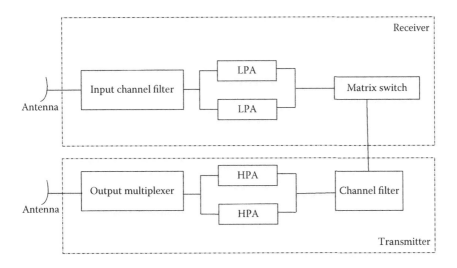

FIGURE 2.11
An example of redundancy in a basic transponder arrangement.

the generation of a set of operational scenarios would have been developed by the system engineers responsible for the operation of each subsystem, which supports the overall system. The test must follow the sequences to be run to test the scenarios.

System engineering evaluation is undertaken by functionally verifying that a requirement or specification is met by observing the qualitative results of an operation or exercise performed under specific or controlled conditions. Consistency of such system performance is the key to maintaining quality control and replication of system units and results.

2.7 Antennas

Based on function, satellite antennas can be classified into the following categories: *communication* and *special satellite antennas* [8]. The communication antennas are used for tracking, telemetry, and command operation throughout mission phases after launch vehicle separation. A nondirectional (i.e., omnidirectional) and circularly polarized antenna is required to ensure the continuous coverage and reception of command signals during mission operation. Two or more broad-beam antennas with switches can be used, instead of a directional antenna, to provide continuous coverage. Special satellite antennas, on the other hand, are directional for specific purposes, such as meteorology, cross-link, earth coverage, and autotracking [12, 17].

Simplistically, an antenna is the interface between a free-space electromagnetic wave and a guided wave. There are many types of antennas and many different variations on the basic types, but their mode of operation is essentially the same. That is, a radiofrequency transmitter excites electric currents in the conductive surface layers of the antenna and radiates an electromagnetic wave. If the same antenna is used with a receiver, the converse process applies; that is, an incident radiowave excites currents in the antenna, which are conducted to the receiver. The ability of an antenna to work both ways is termed the *principle of reciprocity*.

An antenna's radiation varies from omnidirectional to highly directional and can be fixed or changed to accommodate specific needs as they arise. Three basic generic types of antennas are the *reflector, lens,* and *phased array*; however, helical antennas have been deployed in space, particularly for military systems.

A reflector antenna is the most desirable candidate for satellite antennas because of its lightweight, structural simplicity and design maturity. Horns are frequently used as feeds to illuminate reflector antennas, which typically provide narrow beams. A horn antenna can be used alone, usually for wide-angle coverage of the earth. The reflector, however, may need to be offset for a multiple-beam antenna system to avoid feed blockage. The blockage destroys the rotational symmetry of the surface and limits the range scan before aberrations seriously degrade the scanning performance.

A lens antenna is the counterpart of the reflector from the optical property point of view. It can be made rotationally symmetric to preserve good optical characteristics. It has no feed blockage and compactness; however, it is heavier at low-frequency application and has lens surface mismatch.

A phased-array antenna is a class of array antennas that provide beam agility by effecting a progressive change of phase between successive radiators. An antenna array is a family of individual radiators whose characteristics are determined by the geometric position of the radiators and the amplitude and phase of their excitation [18]. A phased-array antenna has a number of advantages over a lens or reflector antenna. This is due to the distribution of power amplification at the elementary radiation level, higher aperture efficiency, no spillover loss, no aperture (feed) blockage, and better reliability.

The helical antenna has inherent broadband properties, possessing desirable pattern, polarization, and impedance characteristics over a relatively wide frequency range and may radiate in several modes.

Each of these antennas is a major study area. However, each is briefly described in the next subsections.

2.7.1 Horn

A horn is a slightly flared end of a piece of waveguide. Instead of electrical currents, the waveguide carries a tightly focused electromagnetic wave with the electric components extending between the parallel walls. Horns can

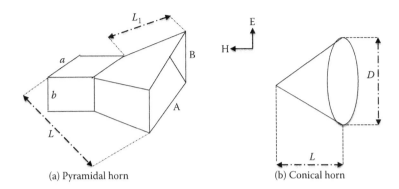

(a) Pyramidal horn   (b) Conical horn

FIGURE 2.12
Horns.

be square in section (called pyramidal horns) but rectangular in either two orthogonal planes (called E-plane and H-plane horns), as in Figure 2.12(a), or circular in cross section, as in Figure 2.12(b).

In the figure, A and B are aperture length in the H-plane and E-plane, respectively; a and b are length and breath of the matching waveguide if the horn's shortest length L_1 possible is required; D is the aperture diameter of the conical horn; and L is the axial length to the apex to the aperture of the conical horn.

Two kinds of horns in common use are the pyramidal and conical horns. A horn's interior surfaces can be smooth or corrugated, depending on polarization requirements. Pyramidal horns are easily designed and often used for earth coverage antennas because of their symmetrical radiation properties. The following equations are applicable to pyramidal horns, whose length is long compared to a wavelength, λ:

$$G = 10\log\eta\,\frac{4\pi}{\lambda^2}\,AB \qquad\qquad \text{dBi} \qquad\qquad (2.38)$$

where G is the gain and η is the pyramidal horn's efficiency, typically 50%. Note that dBi means decibel (isotropic)—being the unit of the forward gain of an antenna compared with a hypothetical isotropic antenna that uniformly distributes energy in all directions. Normally, *linear polarization* is assumed unless otherwise stated.

$$\text{3 dB beamwidth in E-plane: } \theta_E = 54\frac{\lambda}{B} \qquad\qquad \text{deg} \qquad\qquad (2.39)$$

$$\text{3 dB beamwidth in H-plane: } \theta_H = 78\frac{\lambda}{A} \qquad\qquad \text{deg} \qquad\qquad (2.40)$$

If it is necessary to have the shortest length possible, then by scaling

$$L_1 = L\left(1 - \frac{a}{2A} - \frac{b}{2B}\right) \qquad m \qquad (2.41)$$

the gain of the conical horn can be written as

$$G = -2\left(1.41 - 10\log \frac{D}{\lambda}\right) \qquad \text{dBi} \qquad (2.42)$$

If the aperture is uniformly illuminated, the 3-dB beamwidth of a conical horn can be approximated to the 3-dB of a circular aperture of an equivalent illumination:

$$\theta_{BW} \approx 58\frac{\lambda}{D} \qquad \text{deg} \qquad (2.43)$$

The advantage of using conical horns is the ability to exploit higher-mode propagation.

Another variation of the horn feed often used in primary feeds, particularly for big earth stations, is a hybrid-mode horn. When annular corrugations are placed on the inner wall of a circular waveguide, a hybrid-mode horn is formed. If the annular corrugations are placed in such a way that neither transverse electric (TE) nor transverse magnetic (TM) modes can be propagated, then a hybrid mode is generated. The hybrid-mode horn antennas can be used to achieve axially symmetric beamwidths and improve cross-polarization and sidelobe performance.

Example 2.3

The earth subtends an angle of 17.3° when viewed from geostationary orbit. Estimate the dimensions and gain of pyramidal horn and conical horn antennas, which will provide global coverage at 4.5 GHz.

SOLUTION

By assuming a uniformly illuminated wave across the aperture (length and breadth) of the pyramidal antenna, we can take the beamwidths in E- and H-planes to be the same, that is,

$$\theta = 17.3° = \theta_E = \theta_H$$

Take the antenna's efficiency $\eta = 50\%$. Then, the wavelength $\lambda = c/f = 0.3 \times 10^9/4.5 \times 10^9 = 6.67$ cm.

From (2.39) and (2.40), we can compute the aperture dimensions:

A = 30.06 cm
B = 20.81 cm
Gain, G = 19.46 dB

Similarly, the dimensions and gain of the conical can be computed using (2.42) and (2.43):

Aperture diameter, D = 22.35 cm
Gain, G = 17.63 dB

2.7.2 Reflector/Lens Antenna System

The most straightforward design of reflector or lens antennas uses parabolic geometry. The reflector antenna, one of the two most popular for earth station antennas, consists of a reflector—a section of a surface formed by rotating a parabola about its axis—and a feed whose phase is located at the focal point of the paraboloid reflector. This is the main reason the reflector antenna is also called the *prime focus antenna*. The size of the antennas is represented by the diameter D of the reflector (see Figure 2.13). For any value of α, the geometric property of the parabola dictates that

$$fa + ab = \text{constant} \qquad (2.44)$$

The paraboloid reflector is capable of focusing at infinity the electromagnetic rays coming out of its focus. From spherical geometry it is easily recognized that when a primary point-source reflector (also called *feed*) generates spherical wave fronts, they are converted to plane-wave fronts at the antenna aperture. This is consistent with geometric optic approximation, namely, ray tracing obeying Snell's law of reflection. This means that rays emerging

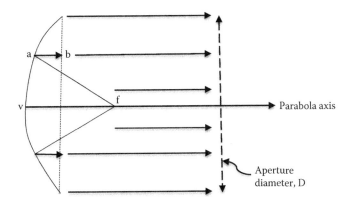

FIGURE 2.13
Geometry of a parabolic antenna, where v = vertex and f = focus of the parabola.

orthogonally (perpendicularly) to the plane-wave front will collimate at infinity, thereby concentrating the radiation in the reflector axis direction. Conversely, when receiving, all the rays intercepted from a plane wave coming from the reflector axis direction will be collimated in the focus, where a suitably located primary radiator will be able to extract maximum energy from the incoming electromagnetic wave. As earlier discussed in terms of the principle of reciprocity, every antenna is a completely reciprocal device, which behaves in exactly the same way (gain, sidelobes, and polarization) in both transmission and reception modes.

The gain and beamwidths of a paraboloidal antenna can be calculated using this formula:

$$G = 10 \log \eta \left(\frac{\pi D}{\lambda} \right)^2 \qquad \text{dBi} \qquad (2.45)$$

The paraboloidal antenna's overall efficiency η is typically within the range 55% to 75%. For design purposes, the mean value is often taken.

Alternatively, if the 3-dB azimuth and 3-dB elevation beamwidths are known, the gain can be written as

$$G = 10 \log \frac{41250\eta}{\theta_e \theta_a} \qquad \text{dBi} \qquad (2.46)$$

where θ_a and θ_e correspond to the 3-dB azimuth beamwidth and 3-dB elevation beamwidth, respectively. Their units are in degrees. If a pencil beam is assumed symmetrical (i.e., $\theta_a = \theta_e$), then the 3-dB beamwidth can be approximated using

$$\theta_a \approx 65 \frac{\lambda}{D\sqrt{\eta}} \qquad \text{deg} \qquad (2.47)$$

Example 2.4

A paraboloid antenna working at 1.8 GHz and having a diameter of 9.5 m is required to achieve a gain of 41.5 dB. Find the necessary efficiency. How much gain variation is associated with ±4.5% efficiency variation?

SOLUTION

From (2.45) the antenna efficiency can be written in terms of other antenna parameters:

$$\eta = 10^{0.1G} \left(\frac{\lambda}{\pi D} \right)^2 = 44.05\%$$

So, for $\eta = 44.05 \pm 4.5\%$, we obtain gain $G = 41.5 \pm 0.4$ dB.

2.7.2.1 Reflector/Lens Antenna Configuration

Many small paraboloidal antennas are front-fed; that is, the feed is situated at the focus of the reflector, as shown in Figure 2.14(a). However, the feeds of transceiver earth station antennas may be bulky, especially if they transmit and receive orthogonally polarized transmissions [8, 9].

There are several variations to the basic parabolic configurations, resulting in the creation of the Cassegrain antenna and Gregorian antenna.

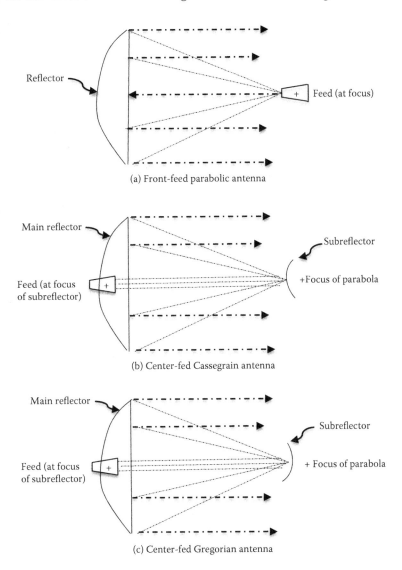

(a) Front-feed parabolic antenna

(b) Center-fed Cassegrain antenna

(c) Center-fed Gregorian antenna

FIGURE 2.14
Axisymmetric reflector antennas.

Cassegrain or Gregorian can be center-fed or offset. The center-fed arrangements are shown in Figure 2.14(b) and (c) for the Cassegrain antenna and Gregorian antenna, respectively. The antennas in Figure 2.14 use main reflectors whose surfaces are symmetrical about the axis of the paraboloid—hence the name *axisymmetric reflectors*. The feeds of these antennas are at one focus of a hyperboloidal subreflector, while the other focus of the hyperboloidal is coincident with the focus of the main reflector.

The advantage of using offset (asymmetric) reflectors is that the subreflector can be put where it does not block or scatter the energy radiated by, or falling on, the antenna aperture. By this arrangement, one can achieve better sidelobe performance than the axisymmetric antennas.

The reflector or lens antenna can be configured to have single-feed or multiple-feed elements. For the multiple-feed elements configuration, as in Figure 2.15, each feed element illuminates the aperture and generates a constituent beam. Any shaped beam can then be formed from a number of these constituent beams by the principle of superposition.

The *principle of superposition* governs the interaction of multiple beams (or waves) passing through an aperture at the same time. For instance, when two beams (or waves) overlap at a point in space, their net displacement is the algebraic sum of the displacements due to each beam (or wave) at that point. The interaction can be *constructive*—the beams (or waves) reinforcing each other—or *destructive*—the beams (or waves) canceling each other.

A reflector or lens antenna with multiple-feed elements is widely used for generating a shaped beam for the satellite communication antenna system. The key features of this antenna system are:

1. Generation of a multibeam pattern from one optical aperture [10]
2. Creation of a steeper pattern roll-off, which results in a higher spatial isolation for the communication system [11]
3. Generation of nulls for an antijam antenna system [9, 12]

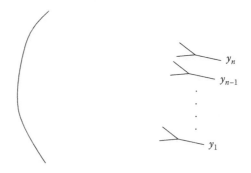

FIGURE 2.15
Feed array elements (y_1, ..., y_{n-1}, y_n) for a multibeam antenna system.

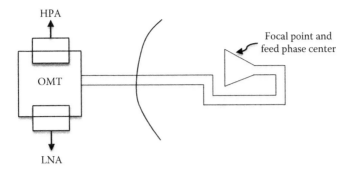

FIGURE 2.16
A transceiver paraboloidal antenna connection with a three-port OMT device.

The feed elements array consists of radiating elements and a beam-forming network, which are either air stripline or waveguide. The selection of the type of elements is determined by the achievable gain, scan loss, cross-polarization, frequency bandwidth, size, and weight.

Nearly all earth station antennas with a diameter greater than 4 m are of the dual-reflector Cassegrain type [12]. The disadvantage of the front-fed configuration, apart from the obvious energy blockage, is that for large antennas a long length of waveguide will be required to carry signals from the focus to the low-noise amplifier. The attenuation of this waveguide raises the noise temperature of the antenna by about 7 K for every 0.1 dB loss. A typical connection of the antenna feed to a *high-power amplifier* (HPA) and *low-noise amplifier* (LNA) through an *orthogonal-mode transducer* (simply *ortho-mode transducer* (OMT)) is shown in Figure 2.16. An OMT is a three-port waveguide device.

A transceiver antenna is a single antenna used for both receiving and transmitting radiowaves. In that instance, receive and transmit signals are typically given opposite polarization to reduce interference and kept separate by an OMT. On the transmit side, the signal energy from the output of the HPA is radiated at the focal point by the feed and illuminates the reflector, which reflects and focuses the signal energy into a narrow beam. On the receive side, the signal energy captured by the reflector converges on the focal point and is received by the feed, which is then routed to the input of the LNA.

2.7.3 Phased-Array Antenna

Array antennas differ physically and electronically from horn and reflector antennas, although several horn elements, forming an array, can be used to generate a shaped beam. The difference results in gain behavior.

The shape of an array antenna can be circular, hexagonal, rectangular, or any geometric shape desired. Figure 2.17 shows some shapes of array antennas.

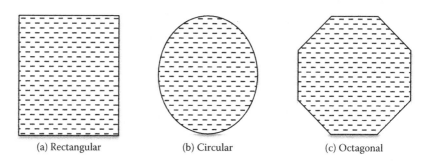

   (a) Rectangular   (b) Circular   (c) Octagonal

FIGURE 2.17
Typical array antenna shapes: (a) rectangular, (b) circular, and (c) octagonal.

The theory of arrays of arbitrary elements is classical and well developed in many texts, including Schekunoff [13], Stratton [14], and Kolawole [18]. By taking the current through the linear array elements to be sinusoidal in time and continuous, we can write the far-field (Fraunhofer zone) radiation pattern as

$$f(\theta) = \left| \frac{\sin\left[\pi \dfrac{D}{\lambda}(\sin\theta - \sin\theta_p) \right]}{\pi \dfrac{D}{\lambda}(\sin\theta - \sin\theta_p)} \right| \tag{2.48}$$

where θ = look angle of the linearly radiating elements; θ_p = steering angle; D = aperture diameter, that is, the physical length of the array; and λ = wavelength in the direction of propagation.

If the steering is in the direction of the look angle, $\theta_p = 0$ and the directive gain can be written as

$$G = 10\log\left(\frac{4\pi A}{\lambda^2} \right) + L_T \qquad \text{dBi} \tag{2.49}$$

where A is the array area and L_T is the total losses, including quantization loss, insertion loss, frequency scan loss, and nonuniform illumination loss. The beamwidth can be estimated upon an application of small-angle approximation giving

$$\theta_{BW} \approx 51\frac{\lambda}{D} \qquad \text{deg} \tag{2.50}$$

The deployment of phased-array antennas for use in a satellite antenna system has been slow primarily due to high manufacturing cost and lack of affordable electronics. Advances in antenna design, materials, production processes, and availability of solid-state electronics have provided a strong incentive to develop a phased-array system for communication satellites. Known deployments of phased-array antennas in satellite communication include the NASA Seasat-A [15], Global-star satellites [16], and Navy geostationary satellites [19, 20]. The use of phased-array antennas on new models of commercial communication satellites is accelerating because of their electronic agility and the capability to be reconfigured in orbit.

Phased-array antennas typically have large numbers of active transmit and receive elements spread across the surface of the antenna panel, instead of a traditional solid or deployable mesh reflector and feed horn arrangement. The other advantages over a lens or reflector are due to the distribution of power amplification at the elementary radiation level, higher aperture efficiency, no spillover loss, no aperture (feed) blockage, and better reliability and the capability to precisely meet specific needs in the spacecraft's area of coverage.

Adaptive phased-array antennas (called *smart antennas*), steerable electronically, are deployed on new-generation satellites, enabling continuous coverage. The main constraint is that the smart antennas will require more sophisticated management of the signal spectrum they provide.

2.7.4 Helical Antenna

A helical antenna has inherent broadband properties, such as possessing desirable pattern, polarization, and impedance characteristics over a relatively wide frequency range and radiating in several modes. Only the axial mode of radiation is described in this section. In this mode, a reflecting ground plane is used to limit radiation to the forward direction. The field is maximum in the direction of the helix axis (Figure 2.18) and is nearly circularly polarized.

The gain, in the axial mode, of a helical antenna can be written as

$$G = 15c_u^2 \frac{N\Delta s}{\lambda^3} \qquad (2.51)$$

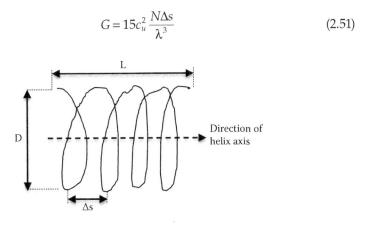

FIGURE 2.18
Dimension of a helical antenna.

where c_u = helix circumference $\approx \pi D$ (m), N = number of turns, Δs = spacing between turns, L = length of helix (m), and λ = wavelength (m).
   Typical limits are:

$$N \geq 3$$

$$0.75\lambda \leq c_u \leq 1.33\lambda$$

$$\Delta s = 0.12\ \lambda;\ \text{spacing from ground plane to first turn}$$

The 3-dB bandwidth of a helical antenna, within these limits, can be written as

$$\theta_{BW} = \frac{52}{c_u}\sqrt{\frac{\lambda^3}{N\Delta s}} \qquad \text{deg} \qquad (2.52)$$

Helical antennas have been deployed on FLTSATCOM satellites to perform transmit and receive functions, having the following dimensions within the wavelength range: 75 cm $\leq \lambda \leq$ 133.3 cm.

   $L = 386$ cm as the receive antenna and mounted in a 76-cm diameter
      cup as its reflecting ground plane
   $L = 340$ cm as the transmit antenna and mounted in a 488-cm diameter
      parabolic reflector as its reflecting ground plane
   $N = 18$ for both configurations

2.8 Satellite Power Systems

A key constraint on satellites is power. Solar or radioactive materials can power the satellite transponder. Due to inherent risk of nuclear fuel, powering by solar energy becomes attractive. The next section looks at the design of a simple power system using solar energy.
   We know that at a distance of 1 astronomical unit (1 AU)—the mean distance between the sun and the earth—the sun delivers an energy flux of about 1370 W/m² (the solar constant). If we are able to harness the sun's energy, we will be able to develop a power source for the satellite. By rule of thumb, the power source may be defined as

$$P_s = A_a \eta_m a_f e_l k_s \qquad (2.53)$$

where P_s = effective solar system power (W), k_s = solar constant = 1370 W/m², A_a = solar array area required for the cells = $l \times b$ (m²), and a_f = loss factor

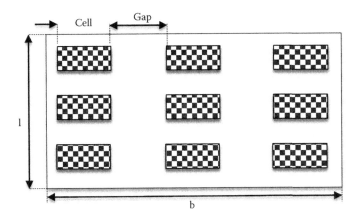

FIGURE 2.19
Plan view of a solar panel.

for uncovered (unused) array area. It is practically impossible to cover all areas of the array panel with solar cells (Figure 2.19). Gaps between cells do not produce power. Also, at certain angles, spacecraft may shadow the sun; hence, this loss factor estimation. A good estimate is about 20% for this loss factor. e_l = electrical losses, which account for losses in panel wiring, and transmission losses in the cell cover glasses. A good estimate is about 20% for electrical losses. η_m = solar cells' conversion efficiency; it depends on the material used. For example:

η_m = 26% for gallium arsenide (GaAs) semiconductor

η_m = 23% for monocrystalline silicon

η_m = 10% for polycrystalline silicon

η_m = 10.6% to 11.4% for dye-sensitized solar cells [21, 22]

The system designer may choose which material to use that will deliver the required power to the system. Research studies are continuing to increase the amount of sunlight solar cells that can be harnessed to about 37% to 40% of the sun's energy. Advanced flywheel technology could provide longer life for power systems and more compact energy storage than chemical battery systems.

During eclipses, solar cells are inert. To keep the communication system operating, backup batteries make up the power system.

Example 2.5

A customer required 140 W continuous power for a communication satellite. Design a power source that meets this requirement.

SOLUTION

You are given $P = 140$ W and solar constant $k_s = 1370$ W/m². You are at liberty to make sensible assumptions such as:

1. Cell material efficiency η_m = say, gallium arsenide (GaAs) semi-conductor, $\eta_m = 26\%$
2. Electrical loss factor $e_l = 20\%$
3. Poorly arranged solar cells leaving gaps a_f uncovered to $\approx 20\%$ of total area

Hence, from (2.53), you require a panel of nominal area $A_a = 9.825$ m².

Example 2.6

A three-array arrangement of solar cells forms the power system of a satellite system. The solar cells are connected in a series/parallel pattern, similar to that shown in Figure 2.19. Preliminary tests indicate that each of the three arrays, with each array having 80 cells, has the following failure rates:

Six short-circuit failures in 10^8 h
Nine open circuit failures in 10^8 h

Calculate: (1) the probability that one solar cell goes (a) short circuit and (b) open circuit in a 1-year period, (2) the probability that the three cells in one arm go short circuit, and (3) the reliability of the solar power system for a 1-year period.

SOLUTION

The solution to Example 2.5 uses the equations derived in (2.32) through (2.36). So,

$N_s = 80$
$dN_f = 6$ for short circuit
$dN_f = 9$ for open circuit
$dt = 10^8$ h
$t = (365 \times 24) = 8760$ h for nonleap year

1. To obtain the probability of the open- and short-circuit cases, each case failure rate and reliability must be established since their probability $P = 1 -$ Reliability. Hence, for:
a. A short-circuit cell, using (2.35), the average failure rate is given by

$$\lambda_{r(s/c)} = \frac{1}{N_s}\frac{dN_f}{dt} = \frac{1}{80}\frac{6}{10^8} = 0.75 {}^* 10^{-9}$$

Using (2.37), the reliability is

$$R = e^{-\lambda_{r(s/c)}t} = 0.99999343$$

Hence, the probability of a short circuit within a calendar year is

$$P_{s/c} = 1 - R_{s/c} = 6.57^* 10^{-6}$$

b. Similarly, for an open-circuit cell:

$$\lambda_{r(o/c)} = \frac{1}{N_s} \frac{dN_f}{dt} = \frac{1}{80} \frac{9}{10^8} = 1.13^* 10^{-9}$$

Its reliability $R_{o/c} = 0.99999015$. The probability of an open circuit within the calendar year $P_{o/c} = (1 - R_{o/c}) = 9.85^*10^{-6}$.

2. The probability of the cell array in one arm short-circuiting is

$$P_{array} = N_s^* P_{s/c} = 80^*6.57^*10^{-6} = 5.26^*10^{-4}$$

3. The probability of both parallel arms short-circuiting is

$$P_{system} = 2^*(P_{s/c})^3 = 80^*6.57^*10^{-6} = 5.26^*10^{-4}$$

Consequently, the reliability of the solar power system over a 1-year period is unity, or 100%.

2.9 Onboard Processing and Switching Systems

New-generation satellite technology requires more sophisticated processing of carrier signals at the satellite prior to retransmission, an area referred to as *onboard processing*. With this processing capability, a more efficient overall satellite link can be realized, and it could lead to a simplification of the present earth stations' requirements. As in the introductory section of this chapter, onboard processing enables the satellite to condition, amplify, or reformat received uplink data and route (or switch) the data to specified locations, or actually regenerate data onboard the spacecraft. The onboard processor can provide a data storage facility required for a switched-beam system and can also perform error correction on uplink data. A satellite system that performs onboard processing is called a *regenerative satellite*.

Onboard processing is divided roughly into two parts: *carrier processing* and *baseband processing*. The basic concept of the carrier processing technique is where the uplink modulated carriers from different sources are distributed

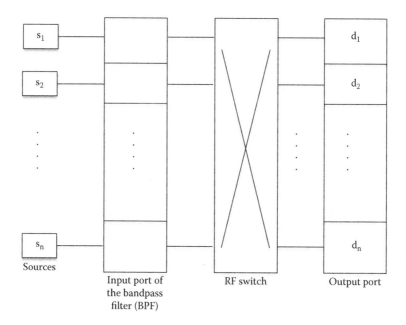

FIGURE 2.20
Carrier processing.

directly to output port (downlink) antennas using fixed or reconfigurable (solid-state) circuitry and switches at the satellite (see Figure 2.20). A switch is reconfigurable when the interconnection patterns among its input and output ports (channel filters) can be changed in response to changing patterns among source-destination pairs. By this technique, no attempt is made to recover the baseband data during the processing, and no changes in modulation formats are allowed.

In the baseband processing, the uplink carriers are first demodulated to baseband data bits (or packets), then demultiplexed, switched (routed), remodulated, and remultiplexed for delivery, each to a destination in different downlink antennas. A schematic diagram of a baseband processor is shown in Figure 2.21. Multiplexing is the ability of multiple users to access a single transponder circuit, also known as the process of combining separate signal channels into one composite stream.

Multiplexing may be done in the time division mode, in which the identity of each source of information is maintained by assigning that information to a particular, periodically recurring time slot of a larger transmission frame (more about time division multiplexing appears in Chapter 5, Section 5.1.2). In a variant of traditional synchronous time division multiplexing, the identity of the individual sources might be maintained by segmenting the informational content of each source into fixed- or variable-length packets, each of which carries a unique identifier within its header field. Of course, the frequency

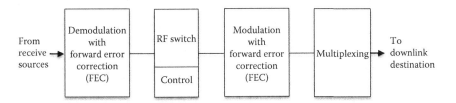

FIGURE 2.21
Baseband processing.

division multiplexing technique can be used as an alternative to time division multiplexing. In frequency division multiplexing, the identity of each source is maintained by modulating that source's information onto a unique carrier frequency, or channel, of a multichannel carrier transmission system (more is said of frequency division multiplexing in Chapter 5, Section 5.1.1).

Demodulation is the process by which the original signal is recovered from a modulated carrier. Demodulation is the reverse of modulation, the process whereby a signal is impressed upon a higher-frequency carrier. This technique modifies baseband signal in a known way to encode information in the baseband signal (more details appear in Chapter 3, Section 3.2.4). Any measurable property of such a signal can be used to transmit information by changing this property in some known manner and then detecting whatever changes at the receiver end. The signal that is modulated is called the *carrier signal* because it carries the digital information from one end of the communication channel to the other end. The device that changes the signal at the transmitting end of the communication channel is called the *modulator*. The device at the receiving end of the channel, which detects the digital information from the modulated signal, is called the *demodulator*. Thermal, device, cosmic, and terrestrial noise and other satellite interference corrupt the satellite channel. It is the function of the demodulators, decoders, and decompressors to recover the signal and produce the desired replica of the transmitted signal. Coders (encoders and decoders) can be stand-alone equipment or can be integrated into a modularized modem: a contraction of *modulation* and *demodulation* (more is said of modems in Chapter 3, Section 3.3.1). The primary function of the *forward error correction* (FEC) coding method is to ensure low error rates over the satellite link. Error correction techniques are one subject of Chapter 6.

2.9.1 Channel Filters and Multiplexers

The purpose of an input channel filter is to confine the bandwidth of the signal allowed into the transponder by rejecting unwanted signals such as those generated by the transponder itself (e.g., spurii) and from other communication systems (e.g., noise and harmonics). Filters are electrical or microwave devices designed to allow a selected range of signal frequencies to pass

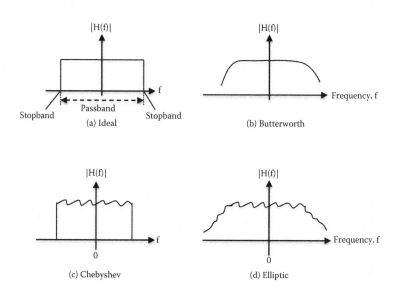

FIGURE 2.22
Type of bandpass filters: (a) ideal, (b) Butterworth, (c) Chebyshev, and (d) elliptic.

while obstructing those outside the range. They also reduce the possibility of interference between transmitted signals.

A generic type of channel filter is the bandpass filter—a filter with both high- and low-frequency cutoffs. The channel filter in communications payload is a bandpass filter since it defines the usable bandwidth of the transponder. A communications payload is divided into two parts: antennas and transponders, parts that have been previously discussed in this chapter. Figure 2.22 shows four types of bandpass filters.

Figure 2.22(a) is a perfectly flat passband filter with an infinite selectivity (or out-of-band rejection). It is purely theoretical. Figure 2.22(b) is the Butterworth filter, which has a flat passband with slow roll-off and poor selectivity. It is rarely used in communications payloads. Figure 2.22(c) is the response of the Chebyshev filter, which solves the selectivity problems at the expense of passband ripple. The ripple is often considered negligible compared with mismatch effects of other equipment. This type of filter is commonly used in communications payload. Figure 2.22(d) shows the response of an elliptic filter: it is a compromise with good selectivity and low ripple but has a limited passband-to-stopband rejection ratio. It tends not to be used as widely as Chebyshev filter for communications payload.

It should be noted that:

The input and output channel filters may be either bandpass or low-pass filters. A low-pass filter has a high-frequency cutoff, allowing only low-frequency signals to pass, whereas a high-pass filter is the opposite of a low-pass filter.

Filters are often installed in the form of input/output multiplexers or diplexers. A *diplexer* is a two-channel multiplexer. *Multiplexers* are devices used to distribute, connect, and combine signals for the amplifiers and antennas. These devices are part of the switching matrix in the communications subsystem of a satellite. In fact, a multiplexer is a multichannel filter. For instance, where a number of channels share the bandwidth available in a satellite transponder, an input multiplexer separates the channel frequencies and routes each carrier to its own amplifier chain. Once amplified, the channels are recombined in an output multiplexer for return transmission.

The channel output filter has a narrower passband than the input filter. It is this output channel filter that defines the *usable bandwidth* of the transponder—an important quantity as far as the service's user is concerned.

2.9.2 Other Onboard Communication Subsystems

For completeness, other components comprising the transponder, depicted in the block diagram in Figure 2.2, are described in this section. The description enriches the reader's understanding of the overall satellite systems.

2.9.2.1 Low-Noise Amplifier (LNA)

The design objective of the LNA is to deliver a good *signal-to-noise ratio* (SNR). The reader will learn in Chapter 4, Section 4.1 that the typical received signal power is extremely low. As such, it is imperative to reject noise introduced in the signal during its passage through the several hundreds or thousands of kilometers of free space. If the noise is carried through the entire chain, it will be successively amplified and finally retransmitted. Two types of LNA are frequently utilized: a bipolar transistor used up to 2 GHz and a field effect transistor (FET) used where carrier frequency is between 2 and 20 GHz.

2.9.2.2 High-Power Amplifier (HPA)

High-power amplifiers are traditionally traveling wave tube amplifiers (TWTAs). The function of the HPA is to provide the radiofrequency power the downlink needs in order to maintain the required *carrier-to-noise ratio* (C/N) at the earth terminals with minimum amplitude variation. The functional relationship between SNR and C/N will become clearer to the reader in Chapter 3. TWTAs are part of a family of microwave tubes, which includes magnetrons and klystrons. These tubes (magnetrons and klystrons) permit frequency agility and in-pulse frequency scanning, which are essential features of modern communication systems. Figure 2.23 shows a klystron.

FIGURE 2.23

A klystron. A klystron is a microwave generator, typically about 1.83 m (6 ft) long and works as follows. The electron gun (1) produces a flow of electrons. The bunching cavities (2) regulate the speed of the electrons so that they arrive in a bunch at the output cavity. The bunch of electrons excites microwaves in the output cavity (3) of the klystron. The microwaves flow into the waveguide (4), which transports them to the accelerator (a device that produces a high-energy high-speed beam of charged particles, e.g., electrons, protons, or heavy ions). The electrons are then absorbed in the beam stop (5). (Courtesy of NASA.)

The operating point chosen for these TWTA tubes is critical because they exhibit nonlinear characteristics above a certain input power, as demonstrated by the typical transfer characteristics of a typical TWTA, shown in Figure 2.24.

As seen in Figure 2.24, at low input levels (i.e., input power ≤ 18 dBm), the output power increases linearly with an increase in the input. Beyond this point, an increase in input power does not produce a corresponding increase in output power; that is, the TWTA begins to saturate. The 1-dB compression point occurs at an input of about −16 dBm. A 1-dB compression point is where

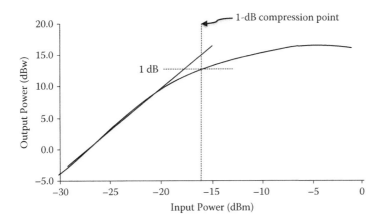

FIGURE 2.24
Typical transfer characteristics of TWTA.

the output power increases 1 dB less than the power increase that would have been obtained from a linear device.

With the advent of phased-array antennas, where each antenna element might have a dedicated high-power amplifier, monolithic microwave integrated circuit (MMIC) solid-state power amplifiers are increasingly being used.

2.9.2.3 Frequency Translator

A frequency translator is required to avoid in-band interference from the high-power satellite output to the satellite input. A frequency translator shifts the uplink frequencies to a different set of downlink frequencies so that some separation exists between the frequency bands in order to prevent unwanted feedback or ringaround, from the downlink antenna into the receiving earth station. In designs where frequency translators are not used, a high isolation between the output and input will be necessary. A typical isolation requirement for a communication system is 27 dB for circular polarization and 33 dB for linear polarization. A frequency translator can be represented by the arrangement in Figure 2.25.

A frequency translator is comprised of a mixer and an oscillator. A mixer combines the incoming signal f_c with the local oscillator frequency f_l prior to being amplified. The mixing process (called heterodyne or heterodyning in short) produces frequencies to the sum and difference of fc and fl. The process is *sum* if the output is of an upconverter and is *difference* if the output is of a downconverter. In the case of communication satellites, the mixer converts the higher-frequency uplink signal to the lower-frequency downlink signal. A local oscillator is a stable frequency source similar to quartz crystal.

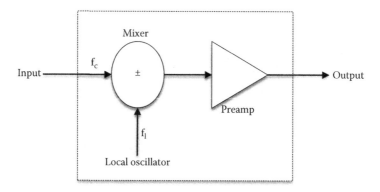

FIGURE 2.25
A simple diagram of a frequency translator.

In general, onboard processing offers a flexible message routing system when various messages have specific destinations, and importantly, it also offers an improved protection against jamming in the military system.

Before concluding this section on satellites, it would be informative to consider another important application area of the satellite: its use in timing is rapidly becoming a critical element for the economy, and positioning. As such, there is need for a technology that can transform navigation and provide precise timing for all users: a global positioning system (GPS) fits this bill.

2.10 GPS

Precise time dissemination is critical to the synchronization of telecommunication networks. Within both wired and wireless systems, consistent pulses and time intervals are used to manage information flow through the network nodes. In particular, the public switched data network (PSDN) relies on accurate timing information for the proper digital transmission of voice and data; wireless telephone and data networks require accurate time to keep all of their base stations in perfect synchronization, thus allowing mobile handsets to share limited radio spectrums more efficiently, and digital broadcast radio services require accurate timing, ensuring that the bits from all radio stations arrive at receivers in lockstep, thus allowing listeners to tune between stations with a minimum of delay.

Precise time is crucial to a variety of economic activities around the world. Besides the communication systems, other networks like electrical power grids and financial networks also rely on precision timing for synchronization and operational efficiency. A GPS provides this critical aspect. Originally, GPS was developed by the U.S. Department of Defense to provide a satellite-based navigation system for its military. It is now used to provide for both military and civilian navigation uses.

A GPS space segment consists of a constellation of satellites transmitting radio signals to users. As noted in Section 2.3.2, a constellation is a group of similar satellites working together in partnership to provide a network of useful services. The GPS constellation is a mix of new and legacy satellites. Currently, there are 32 operational GPS satellites and 4 *residuals* (decommissioned satellites) that can be reactivated if needed. GPS satellites are not in geosynchronous or geostationary orbits. The satellites in the GPS constellation are arranged into properly geometrically spaced 32 slots in eight orbital planes, where each slot contains at least one operational baseline satellite, thus ensuring that there are at least four satellites in view from virtually any point on the planet. Figure 2.26 shows a GPS using a constellation of satellites. The design life of the new-generation GPS satellites is about 7.5 to 11 years; it is likely that a majority would perform well beyond their life expectancy.

Each satellite carries four highly accurate atomic clocks: two cesium (Cs) and two rubidium (Rb), and with in-built selective availability (SA) and antispoof (A-S) capabilities. The GPS satellite's clock has an accuracy of about 20–30 ns. However, because the satellites are constantly moving, the

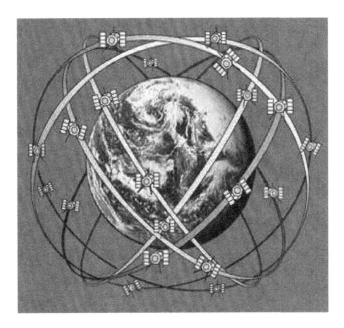

FIGURE 2.26
GPS system using a constellation of satellites. (Courtesy of NASA.)

atomic clocks in the satellites are affected by both special relativity, that is, the satellite's velocity, and general relativity, that is, the difference in the gravitational potential at the satellite's position relative to the potential at the earth's surface [23]. The combinational effect of these two relativities means that the clocks onboard each satellite would tick faster than identical clocks on the ground. The relativistic effects can be computed from a simple mathematical expression that is a function of the semimajor axis, eccentricity, and the eccentric anomaly. Advances in technology have enabled the correction factor being embedded in the GPS receiver and in space GPS satellites. For example, each GPS receiver has an in-built microcomputer that performs the required relativistic calculations when determining the user's location, and the GPS signal in space is expected to provide a worst-case pseudorange accuracy of 7.8 m at a 95% confidence level [24] (more is said about pseuodo-range in Section 2.10.1).

Each GPS satellite transmits a unique navigational signal centered on two L-band frequencies, that is, L_1 at 1575.42 MHz and L_2 at 1227.6 MHz, with enough power to ensure a minimum signal power level of –160 dBw at the earth's surface (the maximum is likely to reach about –153 dBw [25]). Since these microwave signals are highly directional and all GPS satellites transmit on these frequencies, each satellite is assigned individual codes.

Each satellite broadcasts three pseudorandom noise (PRN) ranging codes:

1. The *precision* or *private* (P) code is the principal ranging code and has a far more complex binary sequence, which is approximately 266.4 days long with a chipping rate at the fundamental frequency of 10.23 MHz. The P code is available on both L-band frequencies.

2. The Y code is used in place of the P code whenever the antispoofing (A-S) mode of operation is activated. Under antispoofing, the P code is encrypted through the modulation of the W code, a further secret code. The sum, when encrypted, is called the Y code and is then modulated on the L_1 and L_2 carrier signals. The same P (or Y) code is modulated on both carrier signals, which are then propagated through the ionosphere. The ionosphere exerts some distortion, which in effect *retards* the PRN sequence, but advances the carrier phase. Of course, by line theory, a difference in signal transit time of the same PRN sequence occurs due to the retardation of the two L-band signals by a different amount as they travel through the ionosphere. A detailed treatment of ray refraction through the ionosphere, including estimation of range or time delay error, refraction error, and their cumulative effect on measurements can be seen in Chapter 6 of Kolawole [18].

3. The *clear/access* or *coarse/acquisition* (C/A) code is sometimes also referred to as the S code. The C/A code is used for acquisition of the P (or Y) code (denoted as P(Y)) and as a civil ranging signal.

FIGURE 2.27
GPS control network. (Courtesy of the U.S. National Coordination Office for Space-Based Positioning, Navigation, and Timing, http://www.gps.gov/, accessed May 30, 2013.)

The C/A codes belong to the family of Gold codes* (see brief discussion in Section 2.10.2), which characteristically have low cross-correlation between all members. This property makes it possible to distinguish the signals received simultaneously from different satellites rapidly. The C/A codes are 1023 "chip" long binary sequences that are generated at a 1.023-MHz chip rate used primarily to acquire the P code. The C/A code is available on the L_1 frequency.

Like any complex system, the controlling aspect of the GPS constellation is complex. It comprises a global network of ground facilities, dotted across the globe, as shown in Figure 2.27. The global network tracks the GPS satellites, monitors their transmissions, performs analyses, and sends commands and data to the constellation.

2.10.1 Position Determination

GPS navigation and position determination is based on measuring the distance from the user (or receiver) position to the precise locations of the four GPS satellites as they orbit. If we assume that the distance measured is accurate, then three satellites are sufficient. Suppose, as seen in Figure 2.28, there are three satellites s_1, s_2, and s_3 of known XYZ coordinates (x_1, y_1, z_1), (x_2, y_2, z_2), and (x_3, y_3, z_3), respectively, and unknown user or receiver s_0 coordinate (x_0, y_0, z_0). If the distances between the three satellites' points to the

---

* Including Walsh codes [31], Kasami codes [32], etc.

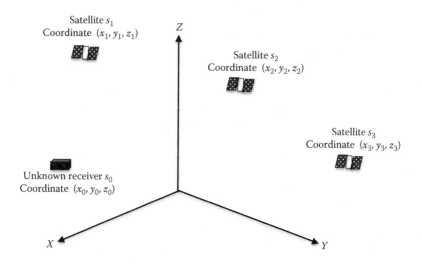

FIGURE 2.28
Use of three known satellites' positions to find an unknown receiver's range, ρ_0.

receiver's point (which is unknown) are measurable, these distances, ρ_1, ρ_2, and ρ_3, can be expressed thus:

$$\rho_1 = \sqrt{(x_1 - x_o)^2 + (y_1 - y_o)^2 + (z_1 - z_o)^2}$$

$$\rho_2 = \sqrt{(x_2 - x_o)^2 + (y_2 - y_o)^2 + (z_2 - z_o)^2} \qquad (2.54)$$

$$\rho_3 = \sqrt{(x_3 - x_o)^2 + (y_3 - y_o)^2 + (z_3 - z_o)^2}$$

The values of x_0, y_0, and z_0 can be determined from three unknowns and three equations. Theoretically, there would occur two sets of solutions, as they are second-order equations, which can be solved by linearization as well as iteratively.

In GPS operation, the positions of the satellites s_1, s_2, and s_3 are given at any time. The distance from the user/receiver (the unknown position) to the satellites must be measured simultaneously at a certain time instance. With the knowledge of time reference associated with each satellite transmitting a signal and measuring the time the signal travels from the satellite to the user, the user's position distance can be found. For instance, suppose the satellite i sends a signal at time t_{si} and the receiver receives it at time t_{ri}, then the distance between the user and the satellite i is

$$\rho_i\big|_T = c(t_{ri} - t_{si}) \qquad (2.55)$$

where c is the speed of light. In Equation (2.55), $\rho_i\big|_T$ is called the *true value of pseudo-range* from user to satellite.

Practically, it is difficult to obtain the exact time from the satellite or the user. The actual satellite clock time t_{si}^a and actual user time t_{ri}^a are related to the true time as

$$t_{si}^a = t_{si} + \delta_i$$

$$t_{ri}^a = t_{ri} + \beta_{ri} \tag{2.56}$$

where δ_i and β_{ri} are satellite clock error and user bias error, respectively. Besides the clock errors, there are other factors, such as:

Satellite position range error, $\Delta\chi_i$
Tropospheric delay error, ΔT_i
Ionospheric delay error, ΔI_i
Receiver measurement error, v_i
Relativistic time correction Δv_i

These affect pseudo-range measurement, causing inaccuracy of the user's position. By factoring Equation (2.56) and these corrections in Equation (2.55), we have the measured pseudo-range ρ_i as [26]

$$\rho_i = \rho_i\big|_T + \Delta\chi_i - c(\delta_i - \beta_{ri}) + c(\Delta T_i + \Delta I_i + \Delta v_i + v_i) \tag{2.57}$$

Some of these errors can be corrected, for example, the tropospheric delay error, which can be modeled, and the ionospheric error by a two-frequency receiver. As noted earlier, advances in technology have enabled a correction factor due to relativistic effects being embedded in the GPS receiver and in space GPS satellites. The user's clock error cannot be corrected through received information, thus remaining an unknown entity.

Errors in estimates of the tropospheric delay have some amount of cancellation relative to GPS positioning because of correlation in the errors for different satellites. Clearly, a constant bias for three pseudo-ranges to three different satellites would not affect position errors at all, but would cause a user clock time bias error. The significant tropospheric errors are likely to be those from satellites at low elevation angles [27].

The positions calculated by a GPS receiver could be in error up to 50 m: the error magnitude will vary with time. Improved GPS uses a ground station to correct the code received from the satellites to within 5 m accuracy. This type of GPS is called differential GPS (DGPS). DGPS uses local radio signals to improve the accuracy of the position ranges calculated by a GPS receiver. By differencing two absolute position determinations, or an application of range corrections to measurements, the improved relative positions (pseudo-ranges) of the DGPS are achieved.

2.10.2 Gold Codes

Codes obtained by adding several maximal-length sequences from separate shift registers of the same length are called *Gold codes*. Gold code [30] sequences are generated by the addition of maximum-length sequences. As an example, suppose a set $G(x, y)$ of Gold sequences is defined using a specified pair of sequences x and y of length l (= $2^n - 1$) for n-stage shift registers. This specified pair is called a *preferred pair*, meaning two specially chosen maximum-length sequences having bounded small cross-correlation within a set.

Cross-correlation between two waveforms X and Y of period T having η duration of 1 bit (chip) is defined as

$$R_{XY}(\tau) = \frac{1}{\eta} \int_T X(t) Y(t + \tau) dt \qquad (2.58)$$

For a pair of sequences x and y of length l to be a preferred pair, the following conditions must be met:

1. n is not divisible by 4. This implies that n is odd or n mod-4 = 2; that is, n yields a remainder of 4 when divided by 2.
2. $y = x[q]$, where q is odd, and $q = 2^k + 1$ or $q = 2^{2k} - 2^k + 1$. (Note that q is the decimation factor.) This expression implies that sequence y is obtained by sampling every qth symbol of x.
3. The greatest common divisor (gcd) of n and k equals:

$$\gcd(n,k) = \begin{cases} 1, & n = 1 \quad \mathrm{mod}\,2 \\ 2, & n = 2 \quad \mathrm{mod}\,4 \end{cases} \qquad (2.59)$$

Now, by considering Figure 2.29 and the preferred pair x and y of length l, we can define a set $G(x, y)$ of Gold sequences by

$$G(x,y) = \{x, y, x \oplus y, x \oplus Ty, x \oplus T^2 y, \ldots, x \oplus T^{l-2} y, x \oplus T^{l-1} y\} \qquad (2.60)$$

where the symbol $\oplus$ denotes addition modulo (mod) 2, and T represents the operator that shifts vectors cyclically to the left by one place. It must be recognized that $G(x, y)$ contains $l + 2$ sequences of period l.

Gold sequences have the property that the cross-correlation between any two, or between shifted versions of them, takes on one of three values:

Value 1 = –1

Value 2 = –$t(n)$

Value 3 = $t(n)$ – 2

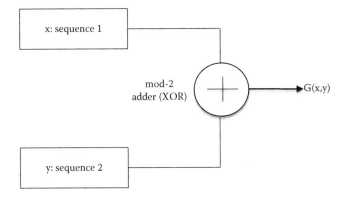

FIGURE 2.29
General structure of preferred pair Gold code generator.

where

$$
t(n) = \begin{cases} 2^{(n+1)/2} + 1 & n = odd \\ 2^{(n+2)/2} + 1 & n = even \end{cases} \tag{2.61}
$$

Example 2.7

Suppose the decimation factor $q = 5$. For a sequence length of 63, test whether preferred pairs are formed to generate Gold code sequences.

SOLUTION

Given $l (= 2^n - 1) = 63$; therefore, $n = \log_2(64) = 6$.
Test the conditionalities of preferred pairs:

1. Since $\mod(n = 6, 4) = 2$, condition 1 is satisfied.
2. If we take $q = 5$, from condition 2, that is, $q = 2^k + 1$ or $q = 2^{2k} - 2^k + 1$, we can obtain the value for $k = 2$. ($5 = 2^k + 1$, and solving for $k = 2$; similarly solve for k using $q = 2^{2k} - 2^k + 1$.) Hence, conditions 2 and 3 are satisfied.

Since the necessary conditions have been met, preferred pairs can be formed to generate a Gold code sequence.

Note that all pairs of l-sequences do not yield Gold codes. However, some pairs of l-sequences with the same degree can be used to generate Gold codes by linearly combining two l-sequences with different offsets in the Galois field—a field that contains a finite number of elements [28]. Of course, Gold code sequences can also be obtained by setting all zeros to the first generator, which is the second l-sequence itself.

L-sequences are often used in *code division multiple-access* (CDMA) systems for spreading and scrambling [17, 29].

2.11 Summary

Communication satellites have evolved from a radio repeater in space to encompass onboard signal processing with switched-beam technology. This chapter has dealt with technical fundamentals for satellite communication services, which do not change as rapidly as technology. The material covered provides the tools necessary for the calculation of basic orbit characteristics such as period, dwell time, and coverage area; antenna system characteristics such as type, size, beamwidth, and aperture-frequency product; and power system design. The system building blocks comprising a satellite transponder and system design procedures have also been addressed.

GPS technology has transformed navigation and precise timing for all users. While GPS is a military-procured and -operated satellite constellation, it has become an essential element of the worldwide economic infrastructure where interdependencies exist between critical infrastructure sectors' operations.

Problems

1. A satellite system consists of 1000 units in series configuration. The failure rate of each unit is 0.48% per 1000 h and is known to be constant. If the mission time of the system is 1000 h, calculate the system reliability. If the desired reliability of the system for a mission time of 1000 h is 0.98, determine what failure rate would be required of the components.

2. An earth station is located at 79.34°W longitude and 37.09°N latitude. Calculate its look angle and range to a geosynchronous satellite whose subsatellite point is located at 102°W longitude.

3. Your colleague has been given a task to design an earth station, which is to be located in your town. As an independent observer, you are encouraged to provide a solution to check your colleague's calculation. The subsatellite point should be within ±22.25° of the

town's longitude. Write a computer program to compute the various look angles and ranges to a satellite located in geosynchronous orbit.

4. A customer required 165.85 W continuous power for a communication satellite. Design a power source that meets this requirement. What additional protection is required to ensure continuity of supply during partial or total eclipse?

5. Given a satellite altitude and desired illumination spot diameter on the earth's surface, simulate the antenna aperture diameter and maximum gain to give the desired spot diameter.

6. Describe briefly the usefulness of an antenna in satellite communication.

7. Describe how the capacity of a transmission channel can be measured.

References

1. Schulz, J.P. (1995). Little LEOs and their launchers. *CommLaw Conspectus*, 3, 185–186.
2. Clarke, A.C. (1945). Extra-terrestrial relays. *Wireless World*.
3. Campbell, D. (2000). Veil drawn around base's role. *The Sunday Age*, July 23.
4. Iridium LCC. (1997). Iridium system facts. http://www.apspg.com/whatsnew/iridium/facts.html.
5. Teledisic. (1998). Technical overview of the Teledesic network. http://www.teledisic.com/tech/details.html.
6. Dalgleish, D.I. (1989). *An introduction to satellite communications*. London: Peter Peregrinus.
7. Hwang, Y. (1992). Satellite antennas. *Proceedings of the IEEE*, 80, 183–193.
8. Kitsuregawa, T. (1990). *Advanced technology in satellite communication antenna: electrical and mechanical design*. Boston: Artech House.
9. Maral, G., and Bousquet, M. (2010). *Satellite communications system: systems, techniques and technology*. New York: John Wiley.
10. Mayhan, J.T. (1983). Adaptive antenna design considerations for satellite communication antennas. *Proceedings of IEEE*, 130, parts F and H.
11. Porcelli, G. (1986). Effects of atmospheric refraction on sun interference. *INTELSAT Technical Memorandum*, IOD-E-86-05.
12. Pritchard, W.L., Suyderhoud, H.G., and Nelson, R.A. (1993). *Satellite communication systems engineering*. Englewood Cliffs, NJ: Prentice-Hall.
13. Schekunoff, S.A. (1943). *Electromagnetic waves*. New York: D. van Nostrand.
14. Stratton, J.A. (1941). *Electromagnetic theory*. New York: McGraw-Hill.
15. *Aviation Week and Space Technology*, December 1977, 51.
16. *Aviation Week and Space Technology*, January 1999, 65.

17. Kolawole, M.O. (2009). *A course in telecommunication engineering*. New Delhi: S. Chand.
18. Kolawole, M.O. (2003). *Radar systems, peak detection and tracking*. Oxford: Elsevier.
19. Rao, J.B.L., Mital, R., Patel, D.P., and Parent, M.G. (2010). *Low cost multibeam phased array antenna for communication with geostationary satellites*. NRL/MR/5310—10-9286. Naval Research Laboratory.
20. Military and Aerospace. 2012, July. http://www.militaryaerospace.com/.
21. Dang, X., Yi, H., Ham, M.-H., Qi, J., Yun, D.S., Ladewski, R., Strano, M.S., Hammond, P.T., and Belcher, A.M. (2011). Virus-templated self-assembled single-walled carbon nanotubes for highly efficient electron collection in photovoltaic devices. *Nature Nanotechnology*, 6, 377–384.
22. BusinessLine. 2012, August. www.thehindubusinessline.com/news/science/.
23. Leick, A. (2004). *GPS satellite surveying*. New York: John Wiley.
24. National Coordination Office for Space-Based Positioning, Navigation, and Timing. http://www.gps.gov/ (accessed May 30, 2013).
25. Navstar (Navigation Satellite Executive Group). (1993). *Global positioning system standard positioning service signal specification*. U.S. Department of Defense.
26. Tsui, J.B.-Y. (2005). *Fundamentals of global positioning system receivers: a software approach*. New York: John Wiley.
27. Spilker, J.J., Axelrad, P., and Enge, P. (1996). *Global positioning system: theory and applications*. Vol. 1. Washington, DC: American Institute of Aeronautics and Astronautics.
28. Calderbank, A.R., Reins, E.M., Shor, P.W., and Sloane, N.J.A. (1998). Quantum error correction via codes over GF(4). *IEEE Transactions on Information Theory*, 44, 1369–1387.
29. Richardson, A. (2005). *WCDMA design handbook*. Cambridge: Cambridge University Press.
30. Gold, R. (1967). Optimal binary sequences for spread spectrum multiplexing. *IEEE Transactions on Information Theory*, 13(5), 619–621.
31. Walsh, J.L. (1923). A closed set of normal orthogonal functions. *American Journal of Mathematics*, 45(1), 5–24.
32. Kasami, T. (1966). *Weight distribution formula for some class of cyclic codes*. Technical Report R-285 (AD632574). Coordinated Science Laboratory, University of Illinois, Urbana.

3

Earth Stations

In Chapter 2, we discussed the basic structure of a satellite system. We now consider a vital part of the overall satellite system: earth stations. An earth station is a collection of equipment for communicating with the satellite, regardless of whether it is a fixed, mobile, aeronautical, or marine terminal. An earth station provides a means of transmitting the modulated radiofrequency (RF) carrier to the satellite within the uplink frequency spectrum and receiving the RF carrier from the satellite with the downlink frequency spectrum. It is the vital link between satellites and communication users.

This chapter considers a basic arrangement of an earth station, the technical and operational requirements, the different types of earth stations used for day-to-day communications, the switching technique employed, the system capacity requirements, and the interface with terrestrial equipment and users.

3.1 Basic Principle of Earth Stations

The basic configuration of an earth station is shown in Figure 3.1. An earth station comprises an antenna, tracking system, receiver, transmitter, multiplexer (combiner), and terrestrial links via a modem (or codec).

Conceptually, the internal electronic signals processing of an earth station is quite simple. For example, let's consider the processing of baseband signals to form the transmitting signals of the earth station. The baseband signals—such as video (TV), voice (telephone), and data—from users are brought via the terrestrial link (e.g., optical fiber, twisted pair cable, coaxial cable, microwave link) from different sources. The baseband signals are then combined (multiplexed)—and encoded with a forward error correction (FEC) scheme in the case of digital satellite communications—and modulated onto the intermediate frequency (IF) carrier to form the earth station's transmission. The transmission can be formatted as a single carrier or multiple contiguous carriers. If the information from a single source is placed on a carrier, the format is called *single channel per carrier* (SCPC). A distinction should be made between a signal and a carrier. A carrier is an electromagnetic wave

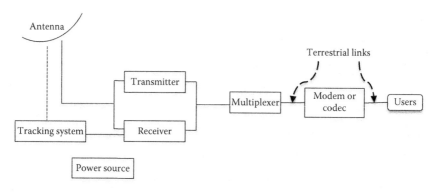

FIGURE 3.1
Block diagram of an earth station.

of fixed amplitude, phase, and frequency that, when modulated by a signal, can be used to convey information through a communication system. The modulated signal can be analog or digital originating from a source, such as a telephone, video, data, and so on, and is often referred to as modulating the carrier. The complete set of these carriers is then radiofrequency translated for high-power amplification and transmission.

If a satellite operates digitally, then analog signals must be converted to digital format before transmission. The conversion is done with a circuit so-called *codec*: a contraction of *code* and *decode* (more is said of codec in Section 3.3.2). The most popular forms of modulation employed in digital communications are binary phase shift keying (BPSK), quadrature phase shift keying (QPSK), offset-quadrature phase shift keying (OQPSK), and 8-PSK. These modulation formats can be described in a general form as M-PSK modulation, where $M = 2^b$ bits, each symbol represents b bits, and b is an integer 1, 2, 3.... The generalized M-PSK is discussed in Section 3.2.

A *modem* converts a digital signal to an analog tone (modulation) and reconverts the analog tone (demodulation) into its original digital signal at the other side of the connection (modems are described in Section 3.3.1). Both amplitude modulation (AM) and frequency modulation (FM) techniques are examples of analog-type or *uncoded* signal transmission systems with one-to-one correspondence between the input signal and the modulated carrier eventually transmitted.

The reader might ask, why code? A coded system makes much more efficient use of bandwidth widening to increase the output signal-to-noise ratio than does an uncoded system. Coded systems are inherently capable of better transmission efficiency than the uncoded types.

A general division may be made to classify frequency division multiplexing (FDM) voice and video signals to use FM while classifying digitally multiplexed voice and data to use some form of phase modulation.

In the receiving chain, the received signals pass through a sequence of processing steps that are essentially the reverse of those followed to prepare the outgoing signals. The RF received signals are passed through a low-noise wideband RF front end, followed by a translator to IF. It is at the IF stage that the specific uplink carriers hoping to be received are separated and grouped by destination, then demodulated to their original baseband signals followed by an FEC decoder (in the case of digital satellite communications). Those not specifically destined to the earth station are retransmitted, thereby acting like a transponder. The demodulated baseband signals may then be demultiplexed, if necessary, and thereby transferred to the users.

The antenna in Figure 3.1 is used for transmitting and receiving: this type is called a transceiver. The basic principles and characteristics of antennas have been discussed in Chapter 2, Section 2.7; they also apply to the earth station antennas. However, in addition to the antenna characteristics discussed in Section 2.7, the antenna subsystem requires separate tracking equipment, which ensures precise pointing of the antenna at the satellite. With small earth stations where the antenna's bandwidth is large, precision tracking equipment is not necessary. The antenna tracking system can be programmed to point to preassigned directions automatically and can also be directed manually. We discuss antenna tracking systems more in Section 3.4.2.

3.1.1 Technical and Operational Requirements

Earth stations are owned and operated either by governments or by private organizations. No matter who the owner is, an earth station must fulfill certain technical requirements in order to meet and maintain system integrity. The technical and operational requirements include the following:

Elimination of any potential interference with local microwave connections, ensuring the coordination of RF bands is in accordance with the International Telecommunication Union (ITU) *Radio Regulations* and relevant local regulations.

Acceptance of the estimated traffic and the performance characteristics of the earth station for each service, over a number of years, which are bilaterally agreeable to the station's owner and satellite providers. Where the provider is the owner, the performance characteristics of the earth station must meet the traffic requirements of its customers and other relevant regulatory requirements.

The earth station must have highly directive gain. This implies that the station's antenna must focus radiated energy into a narrow beam to illuminate the satellite antenna in both the transmit and receive modes in order to provide the required uplink and downlink carrier power.

The earth station must have a low-noise temperature to ensure that the effective noise temperature of its receive side is kept low to reduce the noise power within the downlink carrier bandwidth.

Where a tracking system is employed, the antenna system must be easily steered to ensure that the antenna beam accurately points toward the satellite and minimizes antenna pointing loss.

3.1.2 Types of Earth Stations

Many different types of earth stations are required for satellite communications. The size of an earth station's antenna is the primary feature distinguishing one earth station from another. The types are grouped under three headings: *long earth station, small earth station,* and *very small aperture* (VSAT) *earth station.*

Earth stations with large antennas 10 to 60 m in diameter are called long earth station. This type is often required to provide for high-capacity telephone, data, or television transmission. In general, the larger the antenna, the greater the traffic capacity of the station.

Small earth stations have antennas with diameters between 1 and 10 m. They are commonly sighted on the roofs or in the gardens of domestic and commercial buildings. Small earth stations provide capabilities for reception of broadcast television or connection for thin-route telephony systems in remote regions.

Very small aperture (VSAT) earth stations are networks of satellite earth terminals, each of which has an antenna diameter between 0.3 and 0.9 m—hence the name *very small.* VSAT networks are usually arranged in a star configuration in which small-aperture terminals each communicate via the satellite to a large central earth station known as a *hub* station. Any aperture smaller than VSAT is called an *ultra-small-aperture terminal* (USAT).

Nearly all earth station antennas with diameter greater than 4 m are of the paraboloidal-reflector Cassegrain type (discussed in Chapter 2, Section 2.7.2) with the following advantages [1]:

Greater electrical design flexibility, which allows aperture illumination efficiency to be maximized.

Feed is easily accessible and connectable to LNAs and HPAs.

Primary spillover is normally directed toward clear sky, which picks less noise than the conventional parabolic configurations.

Operational flexibility, particularly in moving the antenna easily in either azimuthal and elevation directions.

3.2 Modulation

Modulation is a signal processing technique that involves switching or keying the amplitude, frequency, or phase of the carrier in accordance with the information binary digits. There are three basic modulation schemes: amplitude shift keying (ASK), frequency shift keying (FSK), and phase shift keying (PSK). These schemes are, respectively, binary equivalent of analog transmission's amplitude modulation (AM), frequency modulation (FM), and phase modulation (PM) when used to transmit data signals.

Consider an input signal $s(t)$ of duration T_s being represented by

$$s(t) = A \cos(2\pi f_c t + \phi_n), \quad (n-1)T_s \leq t \leq nT_s \tag{3.1}$$

where A and ϕ_n correspond to amplitude and nth phase of the signal.

For the ASK scheme, the signal's amplitude A is varied while the phase ϕ_n and carrier frequency f_c remain constant. In the FSK, only the frequency f_c is varied, with A and ϕ_n remaining constant. In the case of PSK, A and f_c are kept constant, with the ϕ_n varied. PSK, compared with the other schemes, has excellent protection against noise because the information is contained within its phase. Noise mainly affects the amplitude of the carrier.

In the M-PSK schemes, the phase of the carrier takes on one of M possible values or symbols. The (stream of) input binary data are first divided into b-bit blocks. Each block is transmitted as one of M possible values or symbols. Each symbol is a carrier frequency sinusoid having one of M possible phase values spaced $2\pi/M$ apart. Then, for the nth M-tuple, say, the M-PSK phase of (3.1) will exist in all intervals:

$$\phi_n \in \left\{ 0, \frac{2\pi}{M}, \frac{4\pi}{M}, \cdots, \frac{2(M-1)\pi}{M} \right\}$$

To ensure that each transmitted binary digit contains an integral number of cycles, the unmodulated carrier frequency $f_c = k_c/T_s$, where k_c is a fixed integer. For example, if the amplitude of (3.1) has a plus or minus unity value, that is, $A = \pm 1$, then $k_c = 2$. A pair of signals that differ in phase by 180° is referred to as *antipodal signals*.

For M-PSK, a symbol duration

$$T_s = T_b \log_2 M$$
$$= bT_b \tag{3.2}$$

providing that there is no reduction in information throughput, where T_b is the input binary interval.

From this M-PSK generalization, one can derive the most important modulation schemes used in satellite communications: BPSK, QPSK or OQPSK, and 8-PSK. The derivation is as follows.

3.2.1 Binary PSK

When the index $b = 1$, $M = 2$, a binary PSK is derived. It is also referred to as BPSK or 2-PSK. In this case, the binary input baseband data $v(t)$ modulates constant amplitude and constant frequency in such a way that two phase values differing by 180° (recall $2\pi/M$, where M in this case is 2) represent the binary symbols 0 and 1, respectively. The corresponding BPSK symbol expression can be written as

$$s(t) = Av(t)\sin\omega_c t \tag{3.3}$$

The BPSK signal constellation (space) diagram is shown in Figure 3.2, where I and Q represent in-phase and in-quadrature axes, respectively.

Instead of the baseband signal $v(t)$ assuming 0 to 1 amplitude values, it could have a *never-return-to-zero* (NRZ) format with ±1 amplitude values, each having period T_s; that is, $T_s = T_b$ from (3.2).

3.2.2 Quadrature PSK

When the index $b = 2$, $M = 4$, a 4-phase PSK is derived. It is also referred to as *quadrature* PSK, or simply QPSK, in which the phase of the carrier in (3.1) is nonzero and can take on one of four values: 45°, 135°, 225°, or 315°. The angular separation between any two adjacent phasors of a QPSK is 90° (recall $2\pi/M$, where M in this case is 4). Figure 3.3 shows the block diagram of a quadrature modulator. Using Figure 3.3, we can consider the

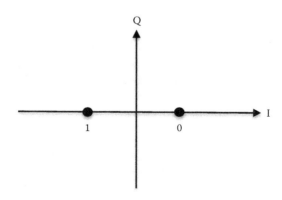

FIGURE 3.2
BPSK space diagram.

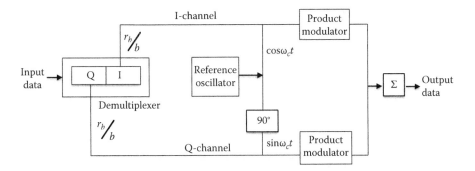

FIGURE 3.3
QPSK modulator.

functions required of QPSK modulation. The incoming serial bit stream enters the bit splitter (demultiplexer), where it is converted to a parallel, two-channel output: I- (in-phase) channel and Q- (in-quadrature) channel. Consequently, the data (bit) rate, r_b, in each of the channels is $r_b/2$, since $b = 2$. The data rate is the signaling speed in which information can be transferred in bits per second (bit/s). The data in each channel are fed to the channel's balanced modulator with the input carrier's relative phase of 0° and 90°, respectively. The outputs of the product modulators are fed to the bit-combining circuit (*linear summer*), where they are converted from the I- and Q-data channels to a single binary output data stream. With $b = 2$, we can code the input bits with corresponding transmitted phase symbols as shown in Table 3.1, with each symbol having a duration of $Ts = 2T_b$. Also, the QPSK constellation (space) diagram is shown in Figure 3.4.

3.2.2.1 OQPSK

Offset QPSK (OQPSK) is a modified form of QPSK. The only difference between QPSK and OQPSK is that the waveforms on I- and Q-channels are shifted or offset in phase from each other by $T_b/2$, that is, one-half of the bit time. Figure 3.5 shows an OQPSK block diagram and the bit alignment

TABLE 3.1

QPSK Truth Table

Input Bits	Transmitted Phase Symbol
00	45°
01	135°
11	225°
10	315°

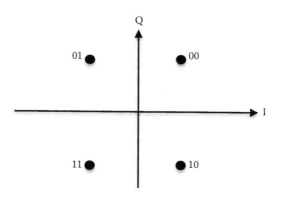

FIGURE 3.4
QPSK space diagram.

diagram. Because changes in the I-channel occur in the midpoints of the Q-channel bits, and vice versa, there is never more than a single bit change in the binary code. As a result, there is never more than a 90° shift in the output phase, which is contrary to the conventional QPSK, where any change in the binary code causes a corresponding 180° shift in the output phase. This phase shift must be imparted during the modulation process. The defect of

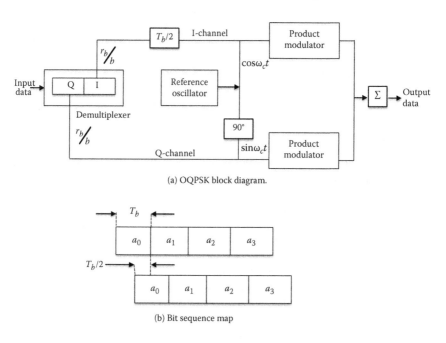

(a) OQPSK block diagram.

(b) Bit sequence map

FIGURE 3.5
OQPSK modulator: (a) OQPSK block diagram and (b) bit sequence map.

TABLE 3.2

8-PSK Truth Table

Input Bits	Transmitted Phase Symbols
111	22.5°
110	67.5°
101	157.5°
100	112.5°
011	337.5°
010	292.5°
001	247.5°
000	202.5°

this modulation scheme is that changes in the output phase only occur at twice the data rate in either channel.

3.2.3 8-PSK

When the index $b = 3$, $M = 8$, an eight-phase PSK is derived. With an 8-PSK modulator, there are eight possible output phases, in which the phase of the carrier in (3.1) can take on one of eight values: 22.5°, 67.5°, 112.5°, 157.5°, 202.5°, 247.5°, 292.5°, or 337.5°, each representing three-input bits as shown in Table 3.2. A schematic diagram of an 8-PSK modulator is shown in Figure 3.6.

The incoming serial bit stream enters the bit splitter (demultiplexer), where it is converted to a parallel, three-channel output: I- (in-phase) channel, Q- (in-quadrature) channel, and the control x-channel. Consequently, the bit rate in each of the channels is $r_b/3$, since $b = 3$. The bits in the I- and Q-channels enter the level (digital-to-analog) converters.

The converted output data are fed to the balanced product modulators with input carriers' relative phase of 0° and 90° to the I- and Q-channels,

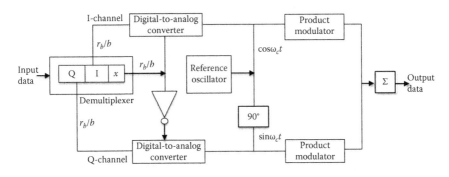

FIGURE 3.6
An 8-PSK modulator.

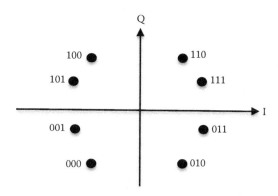

FIGURE 3.7
An 8-PSK space diagram.

respectively. The outputs of the product modulators are fed to the bit-combining circuit (linear summer), where they are converted from the I- and Q-data channels to a single binary output data stream. The 8-QPSK constellation (space) diagram is shown in Figure 3.7.

The angular separation between any two adjacent phasors of an 8-PSK is 45° (recall $2\pi/M$, where M in this case is 8), which is half of what it is with QPSK. This implies that an 8-PSK signal can undergo almost a ±22.5° phase shift during transmission and still retain its integrity.

Higher-order bit encoding techniques (i.e., $b \geq 4$ or $M \geq 16$) can be developed. These techniques are highly susceptible to phase impairments introduced in the transmission medium. If these impairments can be controlled, data can be transmitted at a much higher rate than currently available for the same bandwidth. This imputation becomes obvious in Section 3.2.5.

3.2.4 Demodulation

The demodulation technique reverses the modulation process by converting the analog baseband signals back into a series of digital pulses that the terminal device can use. In general, a coherent demodulator with a decoder is employed (see Figure 3.8) in receiving systems for digital satellite communications.

The decoding operation involves reconstructing the data bit sequence encoded onto the carrier. The lowest possibility of decoding a carrier in error occurs if phase-coherent decoding is used. Phase-coherent decoding requires the decoder to use a referenced carrier at the same frequency and phase as the received modulated carrier during each bit time. In addition, the exact timing of the beginning and ending of each bit must be known. Often the phase and timing coherency is achieved using a separate synchronization device operating together with the decoder. The function of the synchronization device is to extract the required reference and timing from the received modulated carrier itself.

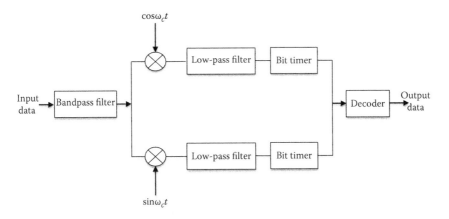

FIGURE 3.8
An *M*-ary PSK demodulator and decoder.

As seen in the previous subsections on PSK techniques, the mapping from binary to *M*-ary symbols is performed in a manner ensuring that any pair of adjacent symbols differs in only one binary digit. Such a mapping scheme is known as *Gray coding* or, sometimes, *maximum distance coding*. This mapping scheme minimizes the number of bit errors that result from a demodulation error because when a digital receiver receives noise-induced error signals, it incorrectly selects a symbol adjacent to the correct one [2]. Using the Gray code results in only a single bit received in error.

3.2.5 Transmission Bandwidth Requirements

Communication speed depends on how fast data are generated from the "source" and transferred through a channel. A channel is the path that propagation signals take. It could be a telephone line, microwave link, satellite link, or optical fiber. A one-way communication path is called a *channel*, and a two-way path formed by two channels is a *circuit*. The data rate is stated in baud. The word *baud* honors the French telegraph operator Emil Baudot, who invented in 1874 a telegraph multiplexer, which allowed signals from up to six different telegraph machines to be transmitted simultaneously over a single wire. *Channel bandwidth* is the frequency band wide enough for, at least, one-way communication.

It can be shown that, for the *M*-PSK system with any value of *M* represented by (3.1), its power spectral density can be expressed as

$$S(f) = A^2 T_s \sin c^2 (\pi T_s [f_c - f_s]) \quad V^2/Hz \tag{3.4}$$

where $f_s = 1/T_s$ (Hz), T_s = symbol interval (s), T_b = input binary interval (s), and A = amplitude of the carrier frequency f_c (V).

Figure 3.9 shows the plot of power spectral density given by (3.4).

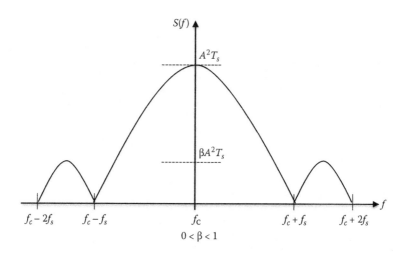

FIGURE 3.9
Unfiltered M-PSK signal power spectral density.

For any M-PSK signals and for a given transmitter frequency, the power spectrum remains the same, independent of the number of symbol levels being used, if the symbol duration T_s is the same. The spectrum efficiency can be written as

$$\eta = \frac{r_b}{B} = \frac{1}{BT_b} \qquad \text{(bit/s-Hz)} \qquad (3.5)$$

where B is the allocated channel bandwidth in Hz. The use of Nyquist filtering (or raised cosine filtering) helps to limit the spectral width without causing adjacent channel interference. If one assumes the filter has a roll-off factor α, then the required channel bandwidth can be expressed as

$$B = \frac{1+\alpha}{T_s} \qquad \text{Hz} \qquad (3.6)$$

By substituting (3.2) in (3.6) and noting that $r_b = 1/T_b$, we can write the maximum bit rate as

$$r_b = \frac{B \log_2 M}{1+\alpha} \qquad \text{bit/s} \qquad (3.7)$$

3.2.6 Probability of Bit Error and Bit Error Rate

In satellite communications it is desirable to minimize the average probability of bit error at the receiver subject to the constraints on received power and channel bandwidth. The terms *probability of bit error* P_e and *bit error rate* (BER) are interchangeably used in the literature, although they differ slightly in meaning in practice. P_e is a mathematical expectation of the BER for a given system. BER is an empirical record of a system's actual bit error performance. In essence, a system performance can be quantified by first measuring the BER and then comparing the BER with the expected probability of bit error P_e.

Probability of bit error is a function of the carrier-to-noise power ratio (C/N). This noise is often taken as thermal noise, which can be expressed as

$$N = kTB \qquad W \qquad (3.8)$$

The carrier power is a function of bit energy and bit duration:

$$C = \frac{E_b}{T_b} \qquad W \qquad (3.9)$$

where k = Boltzmann's constant = 1.38×10^{-23} W/(Hz-K); B = noise bandwidth, Hz; T = system noise temperature, K (note that $0°C \approx 273$ K); and E_b = energy of a single bit, J/bit.

Noise power density N_o is the thermal power N normalized to a 1-Hz bandwidth:

$$N_o = \frac{N}{B} \qquad (3.10)$$

By substituting (3.10) in (3.8) and in view of (3.9), we express

$$\frac{E_b}{N_o} = \frac{CT_b}{N/B}$$

$$= \left(\frac{C}{N}\right)\left(\frac{B}{r_b}\right) \qquad (3.11)$$

where the bit rate $r_b = 1/T_b$. If the bandwidth equals the bit rate, the energy per bit-to-noise power density ratio will equal the carrier-to-noise power ratio, that is, $E_b/N_0 = C/N$.

The general expression for the bit error probability for an M-PSK system is written as

$$P_e = \frac{1}{b} erfc\left(\sqrt{b}\left(\sqrt{\frac{E_b}{N_o}}\right)\sin\left[\frac{\pi}{2^b}\right]\right) \qquad (3.12)$$

where *erfc*(.) denotes the complementary error function. The larger the argument of *erfc*, the smaller is the probability of error P_e.

Error function, *erf*, is by definition

$$erf(u) = \frac{2}{\sqrt{\pi}} \int_0^u e^{-t^2} dt \tag{3.13}$$

and the complementary error function is

$$erfc(u) = 1 - erf(u) \tag{3.14}$$

The integrals for these functions require numerical evaluation. For a large argument *u*, we can use a series expansion of the Laplace-Gauss integral of an asymptotic expansion:

$$\int_0^u e^{-t^2} dt = \frac{\sqrt{\pi}}{2} - \frac{e^{-u^2}}{2u}\left[1 - \frac{1}{2u^2} + \frac{1.3}{(2u^2)^2} - \frac{1.3.5}{(2u^2)^3} + \cdots\right] \tag{3.15}$$

By substituting (3.15) in (3.13) and in turn in (3.14):

$$erfc(u) = \frac{e^{-u^2}}{u\sqrt{\pi}}\left[1 - \frac{1}{2u^2} + \frac{1.3}{(2u^2)^2} - \frac{1.3.5}{(2u^2)^3} + \cdots\right] \tag{3.16}$$

A frequently used function is the *Q* function, which is related to the *erfc* function by

$$Q(u) \cong \frac{1}{\sqrt{2\pi}} \int_u^\infty e^{-x^2/x} dx \tag{3.17a}$$

Comparing (3.17a) with (3.13), we write the relationship between *Q* and *erfc* as

$$Q(u) = \frac{1}{2} erfc\left(\frac{u}{2}\right) \tag{3.17b}$$

For computation purposes, (3.16) in (3.12) enables the *M*-PSK system error rates to be evaluated, and we can observe that P_e decreases monotonically with increasing E_b/N_o. Theoretically, there is a marginal difference between the error performance of the BPSK and QPSK schemes. In practice, however, the error performance of both schemes is taken to be similar because the

3 dB reduction in error distance for QPSK is considered to be offset by the decrease in its bandwidth.

We observe in (3.11) that E_b/N_o is related to carrier power C, input data bit rate r_b, and the channel bandwidth B. If E_b/N_o in (3.12) is replaced with the right-hand terms of (3.11) and the bit rate r_b is kept constant, we must increase the carrier power C to maintain or minimize P_e. This means an increase in the total power capability of the satellite itself or the transmitted power of the earth station. This might not be possible in many satellite systems that are power limited and bandwidth limited. Increasing power or bandwidth is not a solution to minimizing or reducing the probability of error P_e. An effective method of reducing P_e in satellite communications is the use of error correction coding, which is discussed in Chapter 6.

In practice, error performance objectives are usually a compromise between the performance the user wants and what he or she is prepared to pay. Early error performance objectives set by ITU-R (the Radiocommunication Sector of the ITU—see Chapter 7) for an international telephony channel between two earth stations were that BER should be better than:

1 in 10^6 (i.e., 10 min mean value) for at least 80% of any month

1 in 10^4 (i.e., 1 min mean value) for at least 99.7% of any month

1 in 10^3 (i.e., 1 s mean value) for at least 99.99% of any year

For connection via the satellite system, the BER should be better than 1 in 10^3 (i.e., 1 s mean value) for more than 99.97% of the time. A typical objective for a business system (e.g., EUTELSAT SMS systems) is that BER should be greater than 1 in 10^{10} for at least 99% of the year [3]; it is usually necessary to provide error correction in order to achieve this objective. Other objectives for satellite links between network points are available with ITU.

3.2.7 Theoretical Limit on E_b/N_o for an Error-Free Performance

Shannon [4] demonstrated that error introduced by an ideal *additive white Gaussian noise* (AWGN) in a transmission channel of finite bandwidth B can be reduced. Doing so is possible—without sacrificing the rate of information transmission—by encoding information whose transmission bit rate r_b does not exceed the *channel capacity* C_c. The AWGN is assumed to have a spectral density $N_o/2$. Shannon [4] gave the channel capacity as

$$C_c = B \log_2\left(1 + \frac{C}{N}\right) \tag{3.18}$$

In view of (3.11), Equation (3.18) can be written as

$$C_c = B \log_2 \left(1 + \frac{r_b E_b}{B N_o} \right)$$

$$= r_b \frac{E_b}{N_o} \left[\frac{B N_o}{r_b E_b} \log_2 \left(1 + \frac{r_b E_b}{B N_o} \right) \right] \qquad (3.19)$$

By taking the limit of C_c as the channel bandwidth tends to infinity ($B \to \infty$), we can estimate the theoretical limit the channel places on E_b/N_o. By L'Hôpital's rule [5], the limit of a sequence is

$$\lim_{x \to \infty} \left(1 + \frac{1}{x} \right)^x = e \qquad (3.20a)$$

Taking log to base 2 of this expression, we have

$$\lim_{x \to \infty} x \log_2 \left(1 + \frac{1}{x} \right) = \log_2 e$$

$$= \frac{1}{\log_e 2} = 1.443 \qquad (3.20b)$$

noting that

$$\log_2 e = \frac{\log_{10} e}{\log_{10} 2} = \frac{1}{\log_e 2} \text{ and } e = 2.718281828459\ldots$$

By taking the limit of (3.19) and comparing it with (3.20b), we can write

$$\lim_{B \to \infty} C_c = 1.443 r_b \left(\frac{E_b}{N_o} \right) \qquad (3.21)$$

Since r_b must always be less than $\lim_{B \to \infty} C_c$, we can write

$$r_b \leq 1.443 r_b \left(\frac{E_b}{N_o} \right) \qquad (3.22)$$

Alternatively,

$$\frac{E_b}{N_o} \geq \frac{1}{1.443} \geq 0.693 \, (\geq -1.6 \text{ dB}) \qquad (3.23)$$

This is the required minimum value of E_b/N_o, the absolute limit the channel capacity places to communicate with an error-free performance. In other words, a coding scheme exists that allows an arbitrarily low error rate. For example, PSK modulation requires $E_b/N_o \approx 9.6$ dB to achieve a probability of bit error rate P_e of 10^{-5}. Relating this to (3.23) shows that a margin of 11.2 dB is available for error correction coding to achieve the same probability of bit error rate of 10^{-5}.

3.3 Modem and Codec

3.3.1 Modem

The need to communicate between distant computers led to the modem. Modems can operate synchronously or asynchronously. Asynchronous modems operate in the same manner as synchronous modems. However, synchronous modems require at least a scrambler and timer in the transmitter with a matching descrambler and timer at the receiving end. Asynchronous modems use frequency shift keying (FSK) modulation techniques, which use two frequencies for transmission and another two for receiving. No clock is required for asynchronous data transmission or reception because transmitting and receiving modems know only their corresponding nominal data rate. A schematic diagram of a synchronous modem is shown in Figure 3.10.

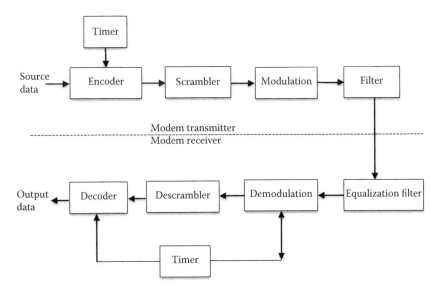

FIGURE 3.10
Basic building block of a synchronous modem.

Basically, a modem consists of a transmitter and receiver, as seen in Figure 3.10, and of course power supply to provide the voltage necessary to operate the modem's circuitry (if the modem is a stand-alone device). A brief functionality of the modem circuitries is described as follows.

3.3.1.1 Encoder and Decoder

An encoder is an option built into many modems. It is used in conjunction with some modulation schemes, enabling each signal to represent more than 1 bit of information. *Encoding* can be either binary encoding (bit by bit) or group encoding (groups of bits). The encoding scheme may be augmented by the forward error correction technique to improve performance.

For a synchronous modem's received timer to function properly, it must remain in synchronization with the data being received. This requires enough changes in the composition of the data—for instance, 0 to 1 and 1 to 0—to permit the receiving modem's circuitry to derive timing from the received data. A clocking signal tells the terminal device when to sample the received data. To prevent slipping of the data relative to the modems' clocks, the data are always grouped in very short blocks (characters) with framing (start and stop) bits. Since the data stream can consist of an arbitrary bit pattern, it is possible for data to randomly contain long sequences of 0s and 1s. When these sequences occur, the data will not provide the modem's receiver with enough transitions for clock recovery, a condition that necessitated the inclusion of scramblers into synchronous modems.

Decoding involves reconstructing the data bit sequence encoded onto the carrier. As indicated in Section 3.2.4, when phase-coherent decoding is used, the lowest possibility of decoding a carrier in error occurs. As such, phase-coherent decoding requires the decoder to use a referenced carrier, possibly from the carrier detector at the same frequency and phase as the received modulated carrier during each bit time. In addition, the exact timing of the beginning and ending of each bit must be known.

3.3.1.2 Scrambler and Descrambler

Scrambling is a coding operation applied to the input information message at the transmitter to randomize the input bit stream. Scrambling eliminates long strings of similar bits that might impair receiver synchronization and periodic bit patterns, which could produce undesirable discrete-frequency components in the power spectrum. In essence, a scrambler modifies the data to be modulated based on a predefined algorithm. A matching descrambler is required at the receiving end so that, after demodulation, the randomized sequence is stripped off. The operation of the scrambler and descrambler is discussed with the aid of Figure 3.11.

The scrambler takes the form of a shift register with feedback connections, while the descrambler is a feedforward connected shift register. A primary

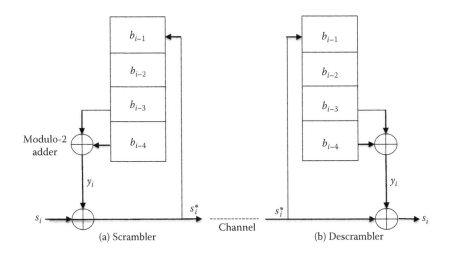

FIGURE 3.11
Principle of scrambling and descrambling: (a) scrambler and (b) descrambler.

reason for the choice of Figure 3.11 is bit error propagation. A single bit error into the feedforward connection affects a successive number of bits equal to the shift register length, while for the feedback connection the effect can be much longer.

Each scrambler and descrambler (in Figure 3.11, for example) employs a four-stage shift register (b_{i-1}, b_{i-2}, b_{i-3}, and b_{i-4}) with appropriate but embedded tap gains and omitting a clock line. The clock clicks whenever a bit sequence enters the register and shifts from one stage to the next. The binary message sequence s_i at the input to the scrambler is mod-2 added to the output of the register y_i to form the scrambled message, which is also fed back to the register input b_{i-1}. Thus, the scrambling operation can be written as

$$y_i = b_{i-3} \oplus b_{i-4}$$

$$s_i^* = s_i \oplus y_i \tag{3.24}$$

The descrambler is a feedforward connected shift register. It essentially reverses the structure of the scrambler and reproduces the original input message sequence because

$$s_i^* \oplus y_i = (s_i \oplus y_i) \oplus y_i$$

$$= s_i \oplus (y_i \oplus y_i) \tag{3.25}$$

$$= s_i \oplus 0 = s_i$$

Equations (3.24) and (3.25) hold for any shift register arrangement as long as the scrambler and descrambler have identical registers.

3.3.1.3 Modulator and Demodulator

A modulator acts on a serial data stream by using the composition of the data to alter the carrier tone that the modem places on the communication line. As expected, the demodulator reverses the process performed by the modulator. When the connection between two modems is established, one modem "raises" a carrier tone that the distant modem hears. The carrier tone by itself conveys no information and is varied by the modulator to impress information that the distant modem demodulates.

When a modem receives a modulated synchronous data stream and passes the demodulated data to an attached terminal device, it also provides a clocking signal to the data terminal. Timing enables data synchronization to be achieved. Synchronization involves sending a group of characters in a continuous bit stream. In the synchronization mode, modems located at each end of the transmission medium normally provide a timing signal or clock to establish the data transmission rate and enable the devices attached to the modems to identify the appropriate characters as they are being transmitted or received. In some instances, the terminal device may provide timing itself or by a communication component, such as a multiplexer or front-end processor channel [6].

No matter what timing source is used, before one begins data transmission, the transmitting and receiving devices must establish synchronization between themselves. To keep the receiving clock in step with the transmitting clock for the duration of a bit stream representing a large number of consecutive characters, the data transmission is preceded by the transmission of one or more special characters. These special characters are designated *syn* in Figure 3.12.

Syn characters are at the same code level as the coded information to be transmitted [7]. However, they have a unique configuration of 0 and 1 bits. Once a group of syn characters is transmitted, the receiver recognizes them and synchronizes itself onto a stream of those syn characters. After synchronization is achieved, the actual data transmission can proceed.

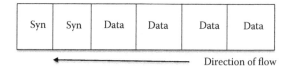

FIGURE 3.12
Data synchronization format.

3.3.1.4 Equalization Filter

An equalization filter (also called simply equalizer) is used to remove a large part of the intersymbol interference introduced by channel filtering [8]. A channel filter, as in Figure 3.10, is used to remove the channel's extraneous tones caused by noise that may have escaped filtering prior to transmission.

To illustrate the basic idea of an equalizer, consider a channel perturbing the received signal, say, $y(t)$, with white Gaussian noise with double-sided power spectral density $(1/2)N_0$. Let's formulate the received signal as

$$y(t) = s_d(t) + \beta_c s_d(t - \tau_\Delta) + n(t) \tag{3.26}$$

where β_c = attenuation due to intersymbol interference; τ_Δ = delay induced by intersymbol interference; $s_d(t) = Ad(t)\cos\omega_c t$, the signal's pattern; $d(t)$ = data stream, a sequence of ±1, each of which is T s in duration; and A = voltage level.

If we take the Fourier transform of this expression and—although important—the noise component $n(t)$ is neglected for simplicity, the transfer function of the channel is

$$H_c(f) = \frac{\Im[y(t)]}{\Im[s_d(t)]} = 1 + \beta_c e^{-j2\pi\tau_\Delta f} \tag{3.27}$$

If β_c and τ_Δ are known, the equalizer will have a transfer function:

$$H_{eq}(f) = \frac{1}{H_c(f)} \tag{3.28}$$

This expression leads to the concept of an inverse filter: a filter best matched to the incoming signal to achieve optimal reception. This type of filter is called a *matched filter*. Since β_c and τ_Δ will not be exact or may even change with time, we must allow for adjusting the parameters of the equalizer. If we restrict ourselves to transversal or tapped delay line structures for the purpose of equalization, we can develop a simple algorithm for estimating the tap weights α_n using either a *zero forcing* or *minimum mean square error* technique. These techniques are described in the next subsections.

3.3.1.4.1 Zero Forcing Technique

Suppose the output of the equalizer $z_{eq}(t)$ in response to the channel output $z_c(t)$ can be written as

$$z_{eq}(t) = \sum_{n=-N}^{N} \alpha_n z_c(t - nT) \tag{3.29}$$

assuming that the tap spacing corresponds to the time T the equalizer is sampled. Our objective is for the equalizer output $z_{eq}(t)$ to satisfy Nyquist's pulse-shaping criterion. So if we put $t = mT$, we can rewrite (3.29) as

$$z_{eq}(mT) = \sum_{n=-N}^{N} \alpha_n z_c[(m-n)T] = \begin{cases} 1, & m = 0 \\ 0, & m \neq 0 \end{cases} \qquad (3.30)$$

where $m = 0, \pm1, \pm2, \pm3, \ldots, \pm N$. Since there will be $2N + 1$ coefficients to be selected from (3.30), it therefore follows that Nyquist's criterion can be satisfied at only $2N + 1$ time instants. If we can write the equalizer output, the tap weights, and channel output in matrix format, Equation (3.30) can be written in the form

$$[z_{eq}] = [h][z_c] \qquad (3.31)$$

where

$$[z_{eq}] = \begin{bmatrix} 0 \\ 0 \\ \vdots \\ 0 \\ 1 \\ 0 \\ 0 \\ \vdots \\ 0 \end{bmatrix} \qquad (3.32a)$$

that is, N zeros at the top and bottom of 1;

$$[h] = \begin{bmatrix} \alpha_{-N} \\ \alpha_{-N+1} \\ \alpha_{-N+2} \\ \vdots \\ \alpha_0 \\ \alpha_1 \\ \vdots \\ \alpha_N \end{bmatrix} \qquad (3.32b)$$

and

$$[z_c] = \begin{bmatrix} z_c(0) & z_c(-T) & \cdots & z_c(-2NT) \\ z_c(T) & z_c(0) & \cdots & z_c[(-2N+1)T] \\ \vdots & \vdots & \cdots & \vdots \\ z_c(2NT) & z_c[(2N-1)T] & \cdots & z_c(0) \end{bmatrix}$$ (3.32c)

The preceding procedure describes the zero forcing technique. Since $[z_{eq}]$ is specified by the zero forcing condition, all that is required is to find $[z_c]^{-1}$, that is, the inverse of $[z_c]$. The desired coefficient matrix $[h]$ is then the middle column of $[z_c]^{-1}$, which follows by multiplying $[z_c]^{-1}$ and $[z_{eq}]$.

As a refresher, the inverse of a nonsingular matrix A is defined as

$$A^{-1} = \frac{A_{adj}}{|A|}$$

where A_{adj} denotes the *adjoint* matrix of A, and $|A|$ denotes the determinant of A. For example, if A is a square matrix, its adjoint is the transpose of the matrix obtained from A by replacing each element of A by its cofactor. This definition becomes clearer with an illustration.

Example 3.1

Find the inverse of matrix A defined by

$$A = \begin{bmatrix} 1 & 2 & 0 \\ 3 & -1 & -2 \\ 1 & 0 & -3 \end{bmatrix}$$ (3.33a)

The adjoint of matrix A is obtained as

$$A_{adj} = \begin{bmatrix} \begin{vmatrix} -1 & -2 \\ 0 & -3 \end{vmatrix} & -\begin{vmatrix} 2 & 0 \\ 0 & -3 \end{vmatrix} & \begin{vmatrix} 2 & 0 \\ -1 & -2 \end{vmatrix} \\ -\begin{vmatrix} 3 & -2 \\ 1 & -3 \end{vmatrix} & \begin{vmatrix} 1 & 0 \\ 1 & -3 \end{vmatrix} & -\begin{vmatrix} 1 & 0 \\ 3 & -2 \end{vmatrix} \\ \begin{vmatrix} 3 & -1 \\ 1 & 0 \end{vmatrix} & -\begin{vmatrix} 1 & 2 \\ 1 & 0 \end{vmatrix} & \begin{vmatrix} 1 & 2 \\ 3 & -1 \end{vmatrix} \end{bmatrix}$$

$$= \begin{bmatrix} 3 & 6 & -4 \\ 7 & -3 & 2 \\ 1 & 2 & -7 \end{bmatrix}$$ (3.33b)

noting that the determinant of

$$C = \begin{bmatrix} a & b & c \\ d & e & f \\ g & h & i \end{bmatrix}$$

$$= a[ei - fh] - b[di - gf] + c[dh - eg] \qquad (3.33c)$$

In view of (3.33c), the determinant of A is

$$|A| = 17 \qquad (3.33d)$$

Hence, the inverse matrix is

$$A^{-1} = \frac{A_{adj}}{|A|} = \begin{bmatrix} \frac{3}{17} & \frac{6}{17} & \frac{-4}{17} \\ \frac{7}{17} & \frac{-3}{17} & \frac{2}{17} \\ \frac{1}{17} & \frac{2}{17} & \frac{-7}{17} \end{bmatrix} \qquad (3.33e)$$

As an exercise for the reader, show that AA_{adj} is a diagonal matrix with diagonal elements equal to $|A|$, that is,

$$AA_{adj} = A_{adj}A = |A|I$$

$$= |A| \begin{bmatrix} 1 & 0 & 0 \\ 0 & 1 & 0 \\ 0 & 0 & 1 \end{bmatrix} \qquad (3.33f)$$

To have a feel for the algorithm described by (3.31), we use Example 3.2 as an illustration.

Example 3.2

Consider a transmission channel for which the following sample values of its pulse response are given:

$$z_c(0) = 1.0 \qquad\qquad z_c(T) = 0.3 \qquad\qquad z_c(2T) = -0.07$$
$$z_c(-2T) = -0.05 \qquad z_c(-T) = 0.2$$

SOLUTION

We can therefore write the channel output matrix as

$$[z_c] = \begin{bmatrix} 1.0 & 0.2 & -0.05 \\ 0.3 & 1.0 & 0.2 \\ -0.07 & 0.3 & 1.0 \end{bmatrix} \qquad (3.34a)$$

and its inverse matrix becomes

$$[z_c]^{-1} = \begin{bmatrix} 1.0815 & -0.2474 & 0.1035 \\ -0.3613 & 1.1465 & -0.2474 \\ 0.1841 & -0.3613 & 1.0815 \end{bmatrix} \qquad (3.34b)$$

The desired coefficient $[\hbar]$ is

$$[\hbar] = [z_c]^{-1}[z_{eq}]$$

$$= \begin{bmatrix} 1.0815 & -0.2474 & 0.1035 \\ -0.3613 & 1.1465 & -0.2474 \\ 0.1841 & -0.3613 & 1.0815 \end{bmatrix} \begin{bmatrix} 0 \\ 1 \\ 0 \end{bmatrix}$$

$$= \begin{bmatrix} -0.2474 \\ 1.1465 \\ -0.3613 \end{bmatrix} \qquad (3.34c)$$

This result is, in effect, the middle column of the inverse matrix that gives the zero forcing tap coefficients. Using these desired coefficients, the equalizer output is thus expressed as

$$z_{eq}(mT) = -0.2474 z_c[(m+1)T] + 1.1465 z_c(mT) - 0.3613 z_c[(m-1)T] \tag{3.34d}$$

The equalizer response of $z_{eq}(mT)$, $m = -1, 0, 1$ at constant sample time $T = 1$, is shown in Figure 3.13.

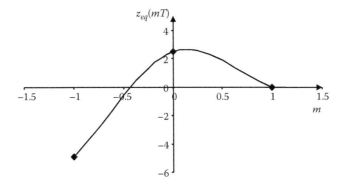

FIGURE 3.13
The three-tap equalizer response using the zero forcing technique.

3.3.1.4.2 Minimum Mean Square Error (MMSE) Technique

The concept of MMSE is to seek the tap weights that minimize the mean square error between the desired output from the equalizer and the actual output at any given time t. We denote the *desired* and *actual* equalizer's output by $z_d(t)$ and $z_a(t)$, respectively. So the MMSE technique is expressed as [3]

$$\varepsilon = E\left\{[z_a(t) - z_d(t)]^2\right\} \tag{3.35}$$

where $E\{.\}$ denotes the expectation of $\{.\}$. Like (3.29), the actual output is

$$z_a(t) = \sum_{n=-N}^{N} \alpha_n z_c(t - n\Delta) \tag{3.36}$$

where, in this instance, the tap spacing Δ does not correspond to the sampling time. Our task is to set a sufficient condition to minimize the tap weights α_n. Since ε is a concave function of α_n, then a sufficient condition for error minimization is

$$\frac{\partial \varepsilon}{\partial \alpha_n} = 0 = 2E\left\{[z_a(t) - z_d(t)]\frac{\partial z_a(t)}{\partial \alpha_n}\right\} \quad n = 0, \pm 1, \dots, \pm N \tag{3.37}$$

Upon substituting (3.36) in (3.37) and differentiating the resultant, we obtain the conditions

$$E\{[z_a(t) - z_d(t)]z_c(t - n\Delta)\} = 0 \quad n = 0, \pm 1, \dots, \pm N \tag{3.38}$$

Alternatively,

$$E\{[z_d(t)]z_c(t - n\Delta)\} = E\{[z_a(t)]z_c(t - n\Delta)\} \quad n = 0, \pm 1, \dots, \pm N \tag{3.39}$$

Let's pause to emphasize what we have obtained in terms of data autocorrelation. By definition, cross-correlation of signals x and y is

$$R_{xy}(\tau) = E\{x(t)y(t + \tau)\} \tag{3.40a}$$

which can be written as

$$R_{xy}(\tau) = E\left\{x(t)\underbrace{[h(t) * x(t + \tau)]}_{y(t+\tau)}\right\} \tag{3.40b}$$

Note that * in (3.40) denotes convolution. So the expectance can still be expanded as

$$E\{x(t)[h(t) * x(t + \tau)]\} = h(\tau) * R_x(\tau) = h(\tau) * E\{x(t)x(t + \tau)\} \qquad (3.41)$$

where R_x is the autocorrelation of x. Using these results, the conditions expressed by (3.39) can be expressed as the autocorrelation function of the equalizer output in terms of the desired data autocorrelation function since only the channel output $z_c(t)$ is fed into the equalizer:

$$[R_{z_d}] = \underset{[h(m)]}{[\hbar]} [R_{z_c}] \qquad (3.42)$$

where $[\hbar]$ = the desired coefficient matrix, the same as defined by (3.32b).

$$[R_{z_c}] = \begin{bmatrix} R_{z_c}(0) & R_{z_c}(\Delta) & \cdots & R_{z_c}(2N\Delta) \\ R_{z_c}(-\Delta) & R_{z_c}(0) & \cdots & R_{z_c}[2(N-1)\Delta] \\ \vdots & \vdots & \cdots & \vdots \\ R_{z_c}(-2N\Delta) & R_{z_c}[(-2N+1)\Delta] & \cdots & R_{z_c}(0) \end{bmatrix} \qquad (3.43a)$$

and

$$[R_{z_d}] = \begin{bmatrix} R_{z_d}(-N\Delta) \\ R_{z_d}[-(N-1)\Delta] \\ \vdots \\ R_{z_d}(N\Delta) \end{bmatrix} \qquad (3.43b)$$

The conditions for obtaining the optimum tap weights using the MMSE criterion are identical to the conditions for the zero forcing weights except that correlation function samples are used instead of pulse response samples.

Two questions readily come to mind regarding setting the tap weights:

1. What should be used for desired response, $z_d(t)$?
2. What procedure should we follow if the sample values are not available?

First, in the case of digital signaling, the detected data should be used. For example, if the modem performance is good, an error probability of 10^{-2} still suggests that the desired output produces bits 99% of the time. Algorithms that use the detected data as the desired output are called *decision directed*.

The answer to (2) is to use the adaptive equalization method, a procedure that allows the initial guess for the coefficients to be corrected according to a recursive relationship

$$\hbar(n+1) = \hbar(n) - \kappa[z_c(n) - z_d(n)]Z_c(n) \qquad (3.44a)$$

where κ = filter weight or adjustment parameter.

$$Z_c(n) = \begin{bmatrix} z_c(n\Delta) \\ z_c[(n-1)\Delta] \\ \vdots \\ z_c(2N\Delta) \end{bmatrix} \qquad (3.44b)$$

The expressions in (3.44) suggest that a training sequence can be periodically sent through the channel, or the equalizer can make use of the incoming data to carry out minimization adjustments.

In summary, the resulting weights for either the zero forcing technique or minimum mean square error technique can be precalculated and preset, or adaptive circuitry can be utilized to automatically adjust the weights.

Modems can be segmented according to their usage by the modulation schemes utilized:

Frequency division multiplexing (FDM) for traffic and video

Single channel per carrier (SCPC) for voice channels' low-level carrier

Time division multiplexing/time division multiple access (TDM/TDMA) for data transmission over satellite links and to data users

Spread spectrum modems, particularly by military satellite communications

The configuration and modulation/demodulation schemes deployed, and their performance characteristics, separate one modem from another. Intersymbol interference introduced by channel filtering can be removed to a large part by zero forcing and minimum mean square error techniques. Carrier acquisition and symbol synchronization or timing extraction are critical to the operation of high-performance modems. With the gradual change to digital transmission, codec became a common feature.

3.3.2 Codec

A codec is an essential part of a communications network. The function of a codec is the reverse of a modem; that is, a codec converts analog input into

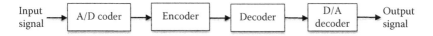

FIGURE 3.14
General structure of a codec.

digital format (A/D), the coder, and digital input analog format (D/A), the decoder. Basically, a codec consists of an analog-to-digital converter (ADC), a digital-to-analog converter (DAC), an internal clock, encoders and decoders, and filtering stages. The basic building block of a codec is shown in Figure 3.14.

The coder portion of Figure 3.14 digitizes the transmitted voice signal, and the decoder reconstructs a received voice signal. In the encoder/decoder units, the digitized signal is decorrelated to reduce redundancies in the signal because a voice signal is usually highly correlated. The filtering stages form part of internal constructs of encoder/decoder units. Their function is to suppress low-frequency noise and apply an antialiasing process to it.

Various types of codecs are available, and their applications vary as the network demands. However, the levels of programmability, the clock frequencies, and the fabrication technologies used dictate the difference. The codec also coordinates the timing between itself and the network with which it interfaces. Data transmission is synchronized and multiplexed. To maintain the industry standard of voice signal quality, the digitized voice signal must have a signal-to-distortion level of at least 30 dB over a 40-dB dynamic range [9]. In order to meet this specification, two methods of data reduction (companding) have been specified: μ-law (also called mu-255 law) and A-law.

Companding, simply a compression and expansion process, is used to upgrade the quality of the pulse, which increases the signal-to-noise ratio and reduces peak power to prevent overloading. Both the encoder and decoder units of the codec perform the companding process.

The μ-law companding technique is the standard used in North America and Japan, whereas the A-law companding technique is the accepted standard in Europe. Most codec manufacturers have developed a device that meets both companding techniques. These laws are logarithmic:

μ-Law:

$$Y = \frac{\ln(1+\mu|x|)}{\ln(1+\mu)} \tag{3.45}$$

where Y = coding range, and x = amplitude, $-1 \le x \le 1$. The value of μ is determined by the law and is constant.

A-Law:

$$Y = \frac{1 + \ln(A|x|)}{1 + \ln(A)} \qquad \text{where} \qquad \frac{1}{A} \le |x| \le 1 \qquad (3.46)$$

$$Y = \frac{Ax}{1 + \ln(A)} \qquad \text{where} \qquad 0 \le |x| \le \frac{1}{A} \qquad (3.47)$$

The value of *A* is determined by the law and is constant.

The main difference between the μ-law and *A*-law is that in *A*-law the normalized amplitude *x* is replaced with a signal whose amplitude is smaller than 1/*A*. Whichever law is used, the prime objective of companding rules is to guarantee a quantization ratio proportional to the size of the original signal.

The most important noise affecting codec design is the idle channel noise, which exists even when there is no signal traffic. This idle channel noise occurs due to the quantization of a 0 V input analog signal as the digital output bounces around the assigned binary value for zero along with any nonlinearity existing along the transmission line.

3.4 Earth Station Design Considerations

Earth stations form an important part of the overall satellite system. In general, when designing an earth station the following must be considered, including implementation restrictions:

1. System parameters such as:

   Transmitter effective isotropic radiated power (EIRP).

   Receiver figure of merit (G/T)—quality of received signal.

   System noise and other interference, including site location, ensuring that the total interference level does not exceed the acceptable level. Interference to, from, and between satellite systems is discussed in Chapter 4, Section 4.2. Other system parameters are interrelated and are discussed in Section 3.4.1.

2. Allowable tracking error relative to satellite drift—upper limit to earth station antenna, addressed in Section 3.4.2.

3. System capacity and availability (covered in Section 3.4.3):

Traffic channel capacity

Service availability and cost—including cost of connection to telecommunications network or the customers' equipment is not excessive.

3.4.1 System Parameters

The earth station's EIRP and quality of received signal (i.e., sensitivity, G/T) measure the earth stations performance. EIRP is the product of the power output P_t of the HPA at the antenna and the gain of the transmitting antenna G_t; that is,

$$EIRP = P_t G_t \qquad (3.48)$$

The receiving system's *sensitivity* is the lowest received signal level for which the system will work without, for example, exceeding the desired error rate or distortion. System sensitivity is often called the *threshold* of the system, which is specified by the receiver antenna gain and the system noise temperature, T. The ratio G/T is called the system *figure of merit*. As demonstrated in Section 2.7, the antenna aperture determines gain. The antenna gains of different antenna configurations were also established in that section. Therefore, the only variable is the system noise temperature T, which is discussed next.

3.4.1.1 System Noise Temperature

Primarily the antenna, the low-noise amplifier (LNA), and the intervening waveguides determine the overall earth station system temperature. This is because the LNA usually reduces the noise contributions from the succeeding devices to a negligible level. We can use Figure 3.15 to express

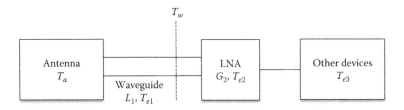

FIGURE 3.15
Block diagram for estimating system noise temperature.

the system temperature. Power loss L_1 rather than power gain G_1 often characterizes the waveguide. Gain is related to the power loss by a simple conversion:

$$G_1 = \frac{1}{L_1} \tag{3.49}$$

In view of (3.8), the noise power at the output of the waveguide can be written as

$$N = \frac{kB}{L_1}(T_a + T_{e1}) \tag{3.50}$$

where T_a and T_{e1} correspond to the antenna noise temperature and waveguide noise temperature. The equivalent noise temperature T_{ex} for a lossy two-port system whose power loss L is greater than unity can be related to the absolute temperature T_o, using the expression

$$T_{ex} = T_o(L - 1) \tag{3.51}$$

Hence, we can write the waveguide noise temperature:

$$T_{e1} = T_o(L_1 - 1) \tag{3.52}$$

Substituting (3.52) in (3.50), we have the noise power at the output of the waveguide:

$$N = kB\left(\frac{T_a}{L_1} + \frac{T_o(L_1 - 1)}{L_1}\right) \tag{3.53}$$

By comparing (3.53) with (3.8), the equivalent noise temperature at the waveguide's output is

$$T_w = \frac{T_a}{L_1} + \frac{T_o(L_1 - 1)}{L_1} \tag{3.54}$$

The earth station system noise temperature referred to the input of the LNA simply is

$$T = T_w + T_e \tag{3.55}$$

where T_e is the equivalent temperature, considering the contributions from LNA and other devices, as seen at the LNA input (at the dotted line point of Figure 3.15):

$$T_e = T_{e2} + \frac{T_{e3}}{G_2} \tag{3.56}$$

By substituting (3.56) and (3.54) in (3.55), we see that the system noise temperature is

$$T = \frac{T_a}{L_1} + \frac{T_o(L_1 - 1)}{L_1} + T_{e2} + \frac{T_{e3}}{G_2} \tag{3.57}$$

With the gain expressions developed for different kinds of antennas in Section 2.7 and the system noise temperature in (3.57), we can establish the earth station sensitivity G/T.

Example 3.3

A 20-m Cassegrain antenna, having a receive antenna gain of 54.53 dB at 11.2 GHz and an ambient temperature of 275 K, is characterized by the following parameters:

Antenna noise temperature = 60 K
Waveguide loss = 0.3 dB
Low-noise amplifier (LNA) with 20 dB gain and 350 K effective noise temperature
Effective noise temperature of the downconverter = 1500 K

Estimate (1) the earth station system noise temperature and (2) the antenna gain referred to the input of the LNA.

SOLUTION

$L_1 = 10^{0.03} = 1.0715$ $\qquad G_2 = 20 \text{ dB} = 10^2 = 100$

$T_a = 60$ $\qquad T_o = 275$ $\qquad T_{e3} = 1500$
$T_{e2} = 350$

1. Using (3.57), we can calculate the earth station temperature as

$$T = \frac{60}{1.0715} + \frac{275 * 0.0715}{1.0715} + 350 + \frac{1500}{100} = 439.345 \text{ K}$$

2. Antenna gain referred to the input of LNA: $G_a = (54.53 - 0.3) = 54.50$ dB.

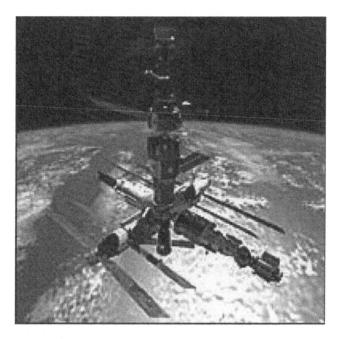

FIGURE 3.16
The Mir space station. (Courtesy of NASA.)

3.4.2 Antenna Tracking

An earth station antenna may be required to work to satellites in different orbital positions during its lifetime. An example is the tracking of the Russian Mir space station (Figure 3.16) as it orbited and guided back to earth on Friday, March 23, 2001. The Mir's progress was charted up to the point of entry. Fragments from the massive complex splashed down in the South Pacific just as ground controllers had planned. It was a technological feat in the history of space tracking and research, and demonstrated the critical importance of precise direction finding.

Some earth station antennas may need to switch between satellites frequently. For example, the antenna may be required to relay programs from a number of different satellite systems to a local broadcasting station or a number of broadcasting stations (like major events, such as military conflicts, the Olympics, or World Cup) or cable TV network. In such cases, it is important to point and repoint the antenna swiftly, simply, and accurately, whether or not it has a tracking system.

Recently deployed satellite communication antennas operate in the 11/14 and 20/30 GHz ranges and have narrower radiation patterns than the older antennas. In order to point the earth station's antenna beam within these narrow sectors of the satellite antenna beam, it is essential to direction-find the satellite with significantly greater precision.

With these examples, how do we track any satellites orbiting the earth and by what method? Three commonly used direction-finding systems in earth stations are monopulse, step track, and programmable steering. The operating principle of all direction-finding systems is based on a comparison of the actual beam axis, aligned in the direction of arrival of signals, with two received radiation patterns: one of the actual beam axis and the other from a satellite (see Figure 3.17).

As seen in the figure, the peaks of the radiation patterns are shifted relative to each other so that the resulting equal-signal line will coincide with the electrical axis of the antenna. When the signals arriving from the direction coinciding with the antenna electrical axis and that received from both radiation patterns have identical strengths, their difference will be zero. As such, no pointing and position error occurs. However, when the direction of arrival deviates from the signal axis, the radiation

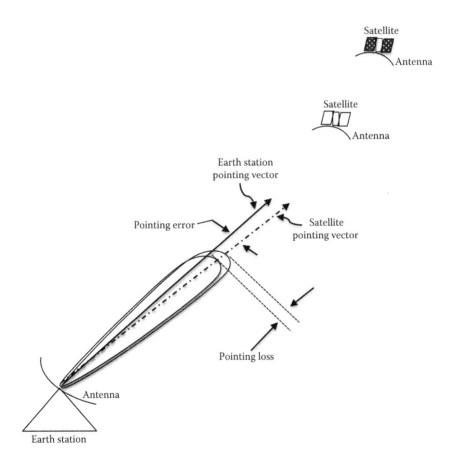

FIGURE 3.17
An illustration of an antenna pointing error.

patterns become nonidentical and the difference results in gain loss or position error.

There are subtle differences in the operation of the three direction-finding systems [10]:

1. Monopulse. In monopulse tracking, multiple-feed elements are used to obtain multiple received signals. The relative signal levels received by the various feed elements are compared to provide azimuth and elevation pointing error signals. The error signals are then used to activate the servo control system controlling antenna pointing. The monopulse method is used in systems that utilize polarization isolation, when greater satellite tracking precision is necessary—for example, in Intelsat antennas.

2. Step track method. In the step track method, the radiation pattern of the antenna is shifted discretely in small steps. The position corresponding to the peak signal is determined by measuring the sign of the difference of the signal levels before and after a step. The differential yields the step size for the next change in position alternately in azimuth and elevation. The iteration process is interrupted when the position is optimum, that is, when the differential is negligibly small. The advantages of the step track method are its simplicity and relatively low cost. Its disadvantage is its low speed.

3. Programmable steering method. In this technique, antenna pointing is based upon knowledge of the relative motion of the satellite with respect to the earth station. A mathematical function, together with the known geographical coordinates of the earth station, is programmed and used to update the antenna pointing without reference to a signal received from the satellite. This technique is independent of the earth station's performance and link parameters, yet it is relatively complex to achieve considerable precision accuracy.

In general, whichever scheme is implemented and in order to prevent the pointing loss, the main lobe of the earth station's antenna must be pointed automatically or manually at the satellite with the greatest possible accuracy. This operation is performed by the tracking system that, by means of its various control loops, ensures that the position errors of the antenna main beam (e.g., due to wind or satellite drift) from the ideal satellite position are compensated. The antenna tracking system together with the tracking servo system, the drive electronics, the electric drives, and the antenna form a closed control loop (as in Figure 3.18). The servo system processes the error signals supplied by the tracking receiver and prepares them for the drive electronics that control the direct current (dc) motors on the antenna axis. The electric drive supplies the drive torque through mechanical gears to the antenna axis and compensates gears' backlash by producing bias torque.

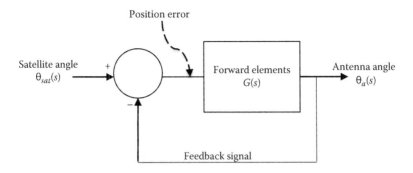

FIGURE 3.18
Tracking feedback control system.

In an earth station's antenna system, a feedback control system with good disturbance rejection as well as good command-following performance is needed. Figure 3.18 depicts a simplified, linear feedback system. Suppose, in this linear feedback control system, the command input and the disturbance are represented by $\theta_{sat}(s)$ and $T_D(s)$, respectively. These functions are related to antenna angle $\theta_a(s)$ by

$$\theta_a(s) = G_c(s)\theta_{sat}(s) + G_D(s)T_D(s) \tag{3.58}$$

For good command-following performance, the transfer function $G_D(s)$ is designed as close to zero as possible. So from (3.58) we write

$$\theta_a(s) = G_c(s)\theta_{sat}(s) \tag{3.59}$$

Alternatively,

$$G_c(s) = \frac{\theta_a(s)}{\theta_{sat}(s)} \tag{3.60}$$

which is the transfer function of the *feedback control system* (of Figure 3.18). This transfer function $G_c(s)$ is related to the open-loop transfer function $G(s)$ by

$$G_c(s) = \frac{G(s)}{1+G(s)} \tag{3.61}$$

where s = Laplace operator. In the analysis of feedback control systems, the Laplace transform is used. The Laplace transform is a transformation from the time t domain to the s domain in an effort to have the frequency response of the signal. Suppose a signal $f(t)$ in the time domain is to be transformed into the s domain, that is, $F(s)$; we can write

$$F(s) = \int_0^\infty f(t)e^{-st}\, dt \qquad (3.62)$$

The signal's frequency response $F(\omega)$ is obtained by substituting s for $j\omega$, where ω is an angular frequency and equals $2\pi f$.

In practice, the transfer function $G_c(s)$ is a product of transfer functions, including that of position compensation element, major and minor rates, and antenna mount [11].

For good command-following performance, the loop gain $G(j\omega)$ of the position feedback loop must be as high as possible while maintaining the stability of the feedback loop despite the mechanical resonance of the antenna mounting, including the reflector assembly.

3.4.3 System Capacity, Cost, and Availability

Like the satellite system availability analysis performed in Section 2.6.5, a similar analysis must be performed for the earth station's system availability. Cost is determined by what the customers want, what level of the performance has been guaranteed, and what is technically achievable. Whether an earth station is owned or leased, it is important that selected equipment meet the operational performance of the satellite communications to be networked to, and that each unit of a particular equipment closely matches the performance of the other. This ensures that the overall system is consistently superior with a high reliability factor, thereby guaranteeing high performance.

3.4.3.1 System Capacity

Video (TV), voice (telephone), and data signals from users via the public switched data networks (e.g., integrated service digital network [ISDN]) are brought through the terrestrial link (e.g., optical fiber, twisted pair cable, coaxial cable, microwave link) from different sources to the earth station. Terrestrial links are investigated in Section 3.5. ISDN basically evolved from telephone networks with an emphasis on multiplexing and switching arrangements, offering a variety of services to a large number of users. ISDN is explored in Chapter 8. Naturally, an earth station may accept traffic from a number of independent terrestrial carriers. The carriers may have been assembled in different access formats (e.g., TDMA, frequency division multiple access [FDMA], or code division multiple access [CDMA]—more about these access formats appears in Chapter 5). In some applications, earth stations may operate in a transponding mode with the same or different assemblies, in which received satellite signals are used to initiate a retransmission from the station to another station or to the satellite. As such, it may be necessary to extract a few channels from each of a number of large assemblies and to disassemble them to the channel level. In this case, the amount of demultiplexing necessary at the earth station may be substantial. The aim of this section is to estimate the channel capacity.

To enhance the reader's understanding of the subject matter, this section briefly introduces information theory and then explains channel capacity analysis. By mastering the basic information theory, the reader will be able to move on to concrete realizations without great difficulty.

3.4.3.2 Information Theory

Information theory deals with the mathematical modeling and analysis of a communication system. It provides limits on:

The minimum number of bits per symbol required to fully represent the information source.

The maximum rate at which reliable communication can take place over the channel, the *Shannon bound*. The Shannon bound provides the motivation for coding.

3.4.3.3 Source Information and Entropy

Figure 3.19 represents a generalized simple model of a digital communication system. The source data contains M symbols from a finite source state x_i, which are traversed via a *discrete memoryless channel* (DMC) to produce data y_i.

The output data y_i, which contains N symbols, can be viewed as a time sequence of source states. The term *memoryless* implies that the source output is independent of all preceding and succeeding source outputs. For convenience, we write the input and output states as

$$X = (x_1, x_2, x_3, \ldots, x_M)$$

$$Y = (y_1, y_2, y_3, \ldots, y_N) \tag{3.63}$$

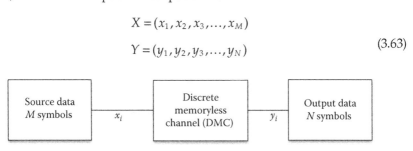

FIGURE 3.19
A generalized model of a digital communication system.

The theory of effective encoding of signals from various sources is a branch of modern information theory. The results of the theory descend from the Shannon theorem, which states that the minimum average number of symbols necessary for data transmission from some sources equals its entropy, H. Thus, entropy implies average information associated with the sources; the average is over all source states. The information associated with source states x_i is

$$I(x_i) = \log_a \frac{1}{p(x_i)} \tag{3.64}$$

where a, the base of logarithm, determines the unit in which the information is measured. Also, $p(x_i)$ is the probability that the source state is transmitted. From elementary mathematics, the sum of the source probabilities is unity:

$$\sum_{i=1}^{M} p(x_i) = 1 \tag{3.65a}$$

Note that

$$\log_a k = \frac{\log_{10} k}{\log_{10} a} \tag{3.65b}$$

Equation (3.64) satisfies these conditions:

1. $I(x_i) \geq 0$ within the bound $0 \leq p(x_i) \leq 1$.
2. As $p(x_i)$ approaches unity, $I(x_i)$ approaches zero.
3. For two independent sources, with states x_r and x_s, the probability of x_r and x_s equals the product of their individual state's probability, $p(x_r)\,p(x_s)$. Consequently, the information conveyed by the joint event of the two symbols is

$$I_{rs} = \log_a \frac{1}{p(x_r)p(x_s)}$$

$$= \log_a \frac{1}{p(x_r)} + \log_a \frac{1}{p(x_s)}$$

$$= I_r + I_s \tag{3.66}$$

This expression shows that the information conveyed by the joint event of the two symbols equals the sum of the information conveyed by each symbol.

Although important, it is not the information associated with source states but rather the average information associated with the source. Hence, the entropy of the DMC source can be written as

$$H(X) = \sum_{i=1}^{M} p(x_i) \log_a \frac{1}{p(x_i)} \tag{3.67}$$

Example 3.4

As an illustration of the meaning of *entropy*, consider a binary source that generates two symbols, 0 and 1, with corresponding probabilities β and $(1 - \beta)$. From (3.67), the binary source entropy is expressed as

$$H(\beta) = \beta \log_2 \frac{1}{\beta} + (1 - \beta) \log_2 \frac{1}{1 - \beta} \tag{3.68}$$

Using this expression, we can draw the entropy function as in Figure 3.20.

Figure 3.20 shows that at the two extremes (i.e., when $\beta = 0$ and 1) the source output is always certain, meaning the entropy is zero; $H(\beta) = 0$. However, if the symbols' probabilities are equally probable—that is, when $\beta = 0.5$—the source output is maximally uncertain and the entropy is maximum, $H(\beta) = 1$. This condition demonstrates that the entropy can be interpreted as a measure of uncertainty.

It is easy to generalize for a DMC with a *binary symmetric channel* (BSC) in which all the M symbols are equally likely to be transmitted and would have an entropy of

$$H(X) = \log_2(M) \tag{3.69}$$

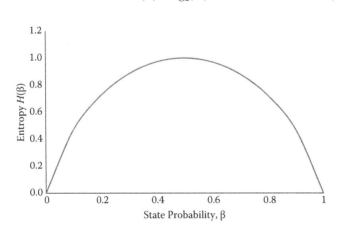

FIGURE 3.20
Entropy of a binary source.

Thus, the entropy could be bounded as

$$0 \le H(X) \le \log_2(M) \tag{3.70}$$

The information transmission rate can be written as

$$R_r = r_b H(X) \tag{3.71}$$

where r_b is the data transmission rate (bit/s).

3.4.3.4 Conditional Probabilities

Where information is mutually shared between the channel's input and output, the entropy measurement will be conditional on the quality of information received either way. The entropy is then interpreted as a conditional measure of uncertainty. Consider a BSC as an example of DMC, as seen in Figure 3.21.

The DMC is defined by a set of transition probabilities. The probability of transmitting x_1 and receiving y_1 is $p(x_1, y_1)$. By Bayes' theorem, this probability can be rewritten as

$$p(x_1, y_1) = p(y_1 \mid x_1)p(x_1) \tag{3.72a}$$

Alternatively,

$$p(x_1, y_1) = p(x_1 \mid y_1)p(y_1) \tag{3.72b}$$

where $p(y_1|x_1)$ = probability of receiving y_1 given that x_1 was transmitted, $p(x_1|y_1)$ = probability of transmitting x_1 given that y_1 was received, $p(x_1)$ = probability of transmitting x_1, and $p(y_1)$ = probability of receiving y_1.

From (3.72), one can write the probability of receiving 0:

$$p(y_1) = p(y_1 \mid x_1)p(x_1) + p(y_1 \mid x_2)p(x_2) \tag{3.73a}$$

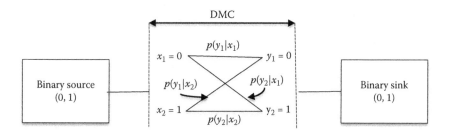

FIGURE 3.21
A representation of a mutual information system using BSC as an example of DMC.

Conversely, the probability of receiving 1 is written as

$$p(y_2) = p(y_2 \mid x_1)p(x_1) + p(y_2 \mid x_2)p(x_2)$$

(3.73b)

Following the preceding procedure, one can generalize for source symbols $(x_1, x_2, x_3, \ldots, x_M)$ and the output symbols $(y_1, y_2, y_3, \ldots, y_N)$. Since the forward transition probability $p(y_j \mid x_i)$ is easy to determine, the receiving probabilities can be written as

$$p(y_j) = p(y_j \mid x_1)p(x_1) + p(y_j \mid x_2)p(x_2) + p(y_j \mid x_3)p(x_3) + \cdots + p(y_j \mid x_M)p(x_M)$$

$$= \sum_{i=1}^{M} p(y_j \mid x_i)p(x_i)$$

(3.74)

The average mutual information $I(X, Y)$ is the average information transferred per symbol across the channel, that is,

$$I(X,Y) = \sum_{i=1}^{M} \sum_{j=1}^{N} p(x_i, y_j) I(x_i, y_j)$$

(3.75)

Comparing this expression with that of the average self-information given by (3.67), the average mutual information becomes

$$I(X,Y) = H(Y) - H(Y \mid X)$$

(3.76)

where $H(Y)$ = output entropy, the average information received per symbol, given by

$$H(Y) = \sum_{j=1}^{N} p(y_j) \log_2 \frac{1}{p(y_j)}$$

(3.77)

$H(Y|X)$ = conditional entropy, the average information lost per symbol, given by

$$H(Y \mid X) = \sum_{i=1}^{M} \sum_{j=1}^{N} p(x_i, y_j) \log_2 \frac{1}{p(y_j \mid x_i)}$$

(3.78)

recognizing that

$$\sum_{j=1}^{N} p(x_i, y_j) = p(x_i)$$

(3.79)

and

$$p(x_i \mid y_i) = \frac{p(x_i, y_i)}{p(y_i)} \tag{3.80}$$

In view of (3.77) to (3.80), the average mutual information of (3.76), may be written as

$$I(X,Y) = \sum_{j=1}^{N} p(y_j) \log_2 \frac{1}{p(y_j)} - \sum_{i=1}^{M} \sum_{j=1}^{N} p(x_i, y_j) \log_2 \frac{1}{p(y_j \mid x_i)} \tag{3.81}$$

By following a similar development, the channel input/output mutual information is

$$I(X,Y) = H(X) - H(X \mid Y) \tag{3.82}$$

Consequently,

$$I(X,Y) = \sum_{i=1}^{M} p(x_i) \log_2 \frac{1}{p(x_i)} - \sum_{i=1}^{M} \sum_{j=1}^{N} p(x_i, y_j) \log_2 \frac{1}{p(x_i \mid y_j)} \tag{3.83}$$

In practice, Equations (3.83) and (3.81) are easier to work with than (3.82) and (3.76) because the channel transition probabilities $p(y_j \mid x_i)$, which are conditional probabilities, can be expressed as a channel matrix:

$$[p(Y \mid X)] = \begin{bmatrix} p(y_1 \mid x_1) & p(y_2 \mid x_1) & \cdots & p(y_N \mid x_1) \\ p(y_1 \mid x_2) & p(y_2 \mid x_2) & \cdots & p(y_N \mid x_2) \\ \vdots & \vdots & \vdots & \vdots \\ p(y_1 \mid x_M) & p(y_2 \mid x_M) & \cdots & p(y_N \mid x_M) \end{bmatrix} \tag{3.84}$$

The generalized transition diagram of the channel can be drawn as shown in Figure 3.22.

In general, the overall conditional probability of a system with many distinct units is the product of individual units' conditional probabilities. For instance, a satellite communication system comprising three channels—the uplink (*u*), satellite transponder processing unit (*s*), and downlink (*d*)—would have its overall conditional probability written as

$$[P(Y \mid X)] = [P(Y \mid X)]_u [P(Y \mid X)]_s [P(Y \mid X)]_d \tag{3.85}$$

To have a feel for this expression, an illustration in the form of Example 3.5 will be examined after the concept of channel capacity has been introduced.

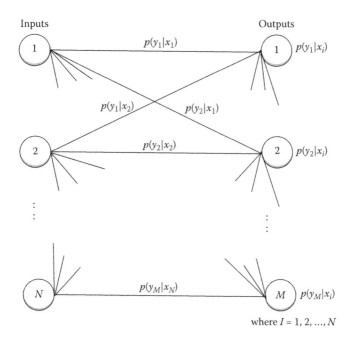

FIGURE 3.22
Generalized DMC transition diagram.

3.4.3.5 Channel Capacity

Channel capacity is the maximum value of mutual information $I(Y, X)$. As an illustration, the BSC as the DMC in Figure 3.21 is used to evaluate the channel capacity. The average information transferred is given by (3.76). The output entropy is

$$H(Y) = \sum_{j=1}^{N=2} p(y_j) \log_2 \frac{1}{p(y_j)}$$

$$= p(y_1) \log_2 \frac{1}{p(y_1)} + p(y_2) \log_2 \frac{1}{p(y_2)} \tag{3.86}$$

By allowing the crossover probabilities to be equal, $(p(y_1|x_2) = p(y_2|x_1) = \alpha)$, and the states probabilities to be $p(x_1) = \beta$ and $p(x_2) = (1 - \beta)$, we can write the receiving self-probability using (3.73) as

$$p(y_1) = (1 - \alpha)\beta + \alpha(1 - \beta) \tag{3.87}$$

Therefore, $p(y_2) = 1 - p(y_1)$. So, the output entropy becomes

$$H(Y) = \Omega(p(y_1)) = \Omega(\beta + \alpha - 2\alpha\beta) \tag{3.88}$$

Following the iterative approach, we can show that the conditional output entropy is

$$H(Y \mid X) = \Omega(\alpha) \tag{3.89}$$

where $\Omega()$ is a functional abstract term. This expression depends on the noise in the channel. Substituting (3.88) and (3.89) in (3.76), the BSC average mutual information is

$$I(X, Y) = \Omega(\beta + \alpha - 2\alpha\beta) - \Omega(\alpha) \tag{3.90}$$

In real life a channel is fixed and the channel's noise is difficult to control. The objectives, therefore, will be to maximize the first $\Omega()$ term, the nonnoise component term, on the right-hand side of (3.90). It was noticed in Figure 3.20 that the entropy $H(\beta)$ is maximum when the symbols' source probabilities are equally probable, that is, $\beta = 0.5$. Therefore, we want the nonnoise component of (3.90) to be

$$0.5 = (\beta + \alpha - 2\alpha\beta) \tag{3.91}$$

This condition is always satisfied when $\beta = 0.5$. This means that if the source encoder can make $\beta = 0.5$ for a BSC, then:

1. Channel capacity $C_c = 1 - \Omega(\alpha)$ bit/symbol $\hfill$ (3.92)
2. Source information rate $R_r = r_b\Omega(\beta)$ bit/s $\hfill$ (3.93a)
3. If $\beta = 0.5$, $R_r = r_b$ (bit/s) $\hfill$ (3.93b)
4. The maximum rate at which information can be transferred across a BSC channel is

$$C_m = r_b C_c = [1 - \Omega(\alpha)]r_b \text{ bit/s} \tag{3.94}$$

Some books refer to this expression as the channel capacity. In practice, Equation (3.92) is often referred to as the channel capacity.

In general, in calculating the channel capacity, whether or not the input data transfer rate is specified, $\Omega(\alpha)$ is replaced with each channel's self-entropy, $H(X)$.

Example 3.5

Consider a digital satellite system to comprise three channels: the uplink, onboard processor, and downlink. The uplink has an error probability of 0.001. The processing unit has an error probability of 0.00025. The downlink error probability is 0.01. Calculate:

1. Each channel's capacity
2. The system's overall conditional probability
3. The overall system channel capacity

SOLUTION

1. Define each channel's error probability $p(y_1)$, $p(y_2) = 1 - p(y_1)$. Using (3.85) and (3.92), we can calculate each channel as

$$C_c = 1 - \left[p(y_1)\log_2 \frac{1}{p(y_1)} + p(y_2)\log_2 \frac{1}{p(y_2)} \right] \text{ bits/symbol} \qquad (3.95)$$

Thus:

Channels	$p(y_1)$	$p(y_2)$	Capacity, C_c
Uplink	0.001	0.999	0.9886
Onboard processor	0.00025	0.99975	0.9966
Downlink	0.01	0.99	0.9192

2. The system's overall conditional probability, using (3.84), can be arranged thus:

$$[P(X\,|\,Y)] = \begin{bmatrix} 0.999 & 0.001 \\ 0.001 & 0.999 \end{bmatrix} \begin{bmatrix} 0.99975 & 0.00025 \\ 0.00025 & 0.99975 \end{bmatrix} \begin{bmatrix} 0.99 & 0.01 \\ 0.01 & 0.99 \end{bmatrix}$$

$$= \begin{bmatrix} 0.9878 & 0.0112 \\ 0.0112 & 0.9878 \end{bmatrix}$$

3. Since the conditional probability results of part (2) are symmetrical, the overall system capacity can be calculated using (3.95):

$$C_c = 0.91259 \quad \text{(bit/symbol)}$$

It should be noted that in the event that $[P(X\,|\,Y)]$ is not symmetrical, select the elements with higher values to calculate the overall system capacity. For example, let the 2×2 elements of $[P(X\,|\,Y)]$ be denoted by α_{11}, α_{12}, α_{21}, and α_{22}. If $\alpha_{11} \neq \alpha_{22}$ and $\alpha_{12} \neq \alpha_{21}$, then:

Select α_{11} if $\alpha_{11} > \alpha_{22}$.
Select α_{12} if $\alpha_{12} > \alpha_{21}$.

3.5 Terrestrial Links from and to Earth Stations

Terrestrial links are primarily optical fibers, microwave, and coaxial cables. The most favored use in digital transmission is optical links because their available broad bandwidths open the way to high data transmission rates, with minimum error probabilities. Optical and microwave links are discussed in this section.

3.5.1 Optical Fiber

Optical fibers are extensively used in telecommunications to transfer information over distance. Their function is to act as a guide for light waves. Optical fibers are glass hairs that are as tough as metal wires but lighter. The diameter of the core that carries the light is only 10 microns (10 millionths of a meter). Fibers are packaged into cables that can be strung from poles, buried in the ground, or carried in ducts; see Figure 3.23(a).

How does an optical link function? An optoelectrical converter takes electrical signals from the user's device (e.g., telephone, computer, or paging or fax machines) and turns them into light pulses. After conversion, light is transmitted down the fiber as digital pulses, usually from a semiconductor laser.

Light then propagates in a step index optical fiber through its internal reflections. That is, as the light pulses travel along the fiber's inner core, they hit the surface of the cladding glass and bounce back, in a series of glancing collisions. Using laser diodes, the propagated burst of light is to indicate binary digit 1 and no light indicates binary digit 0. The light input is then modulated by a scheme known as *optical on-off keying* (OOK). At the other end, a light-sensitive

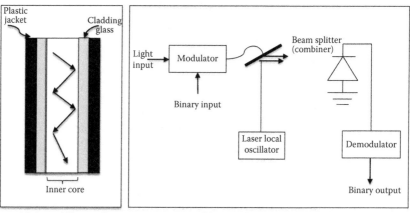

(a) Optical fiber packaging　　　　　(b) Optical circuit

FIGURE 3.23
Optical fiber.

laser diode, for instance, detects the pulses and turns them back to electrical signals for sending to another terminal (see Figure 3.23(b)).

The different angles of rays entering an optical fiber are called propagation modes. Different propagation modes have different speeds of propagation along the fiber. Only a finite set of modes can propagate along an optical fiber. If the radius of the optical fiber is small enough, then only one mode can propagate (i.e., single-mode fibers). Propagation in optical fibers is subject to multimode pulse spreading, attenuation, and dispersion. These effects limit the transmission rate and usable length of a fiber. This is why optical fiber links sometimes require repeaters every 30 to 50 km. Encouraging research has shown how inline optical signal amplification is possible, thereby making the receivers more sensitive, since such signals are weak and quickly overcome by noise, particularly thermal noise.

With wavelength multiplexing—a term used to describe the combined transmission of multiple optical signals of different wavelengths over a common optical fiber—optical fiber links have shown a larger capacity to carry traffic. An example is the *dense wavelength division multiplexing* (DWDM) system, which uses light of different colors traveling down an optical fiber cable.

3.5.2 Microwave

In a microwave radio system, telecommunication traffic is transmitted in the form of direct beams of microwave energy. Microwaves, like light, travel in a straight line between antenna dishes that concentrate the beam. In some cases, the microwave dishes are used as part of a transmission network; they act as *repeater stations*, which receive, amplify, and retransmit the signal they collect. These repeater stations are located at regular intervals in such a way that the overall transmission route is slightly zigzagged to account for the earth's curvature and, importantly, to prevent signals passing between two repeaters overshooting into subsequent paths. In this way signal level is relatively maintained. At the distant end, the receiver detects the signal level, which a multiplexer delivers to the end if required.

3.6 Summary

Earth stations form a vital part of the overall satellite system. This chapter discusses many types of earth stations. In some applications, earth stations operate in a transponding mode, in which received satellite signals are used to initiate a retransmission from the station to another or to the satellite.

For digital transmission, one of the most important performance criteria is the bit error rate. This rate depends, in part, on the type of modulation scheme used for transmission. This chapter also examines the *M*-PSK modulation scheme and develops some useful expressions for calculating the bit error rate, transmission bandwidth, and absolute limit to communicate with an error-free performance.

System noise temperature is one of the important variables required in evaluating the performance of an earth station. The text describes a method that calculates this variable. In order to prevent antenna pointing loss, the main lobe of the earth station's antenna must be pointed automatically or manually at the satellite with the greatest possible accuracy. The commonly used methods for achieving pointing accuracy are monopulse, step track, and programmable steering. Finally, system capacity requirements, as well as the primary terrestrial links to and from the earth station, that facilitate data transmissions are discussed.

Problems

1. Your boss named you the system analyst to integrate the company's new earth station in order to access an existing satellite network, for example, INTELSAT or EUTELSAT. Describe the factors you will consider to implement this task.

2. An M-ary communication system transmits data at a rate of 10^4 symbols/s. Estimate the equivalent bit rates in bit/s for $M = 4, 8, 64$.

3. BPSK and QPSK systems are designed to transmit equal (a) rates (that is, two bits are transmitted with the BPSK system for each phase in the QPSK system) and (b) transmission bandwidths. Compare the systems' symbol error probabilities vs. energy per bit-to-noise power density ratio.

4. An earth station received a digital signal from a satellite with an energy per bit-to-noise power density ratio of 20 dB.

    a. If the required bit error rate is 10^{-5}, what is the maximum data transmission rate possible with a PSK modulation?

    b. For an available bandwidth of 100 Hz, compare your result with the capacity of the system. What is the required energy per bit-to-noise power density ratio at capacity?

5. You will investigate digital communication from a geosynchronous satellite to an earth station. The following parameters are given:

    Transmission power = 9.5 W.

    Frequency = 0.35 GHz.

    Antenna gains of the satellite and earth station are 10.05 and 29.75 dB, respectively.

Additive noise spectral density is 10^{-20}W/Hz.

Desirable error probability is 10^{-5}.

a. What is the maximum data rate the system will allow, assuming an ideal PSK modulation?

b. Find the maximum binary rate for $M = 32$.

c. The error probability must be extremely small. What is the maximum transmission rate possible if an arbitrarily large bandwidth and complex encoding scheme is allowed?

6. Conventional BPSK modulation allows the transmission of either $0°$ or $180°$ phase values. If the phases in the signal are $60°$ and $120°$, will the resulting waveform still remain BPSK?

7. If data are scrambled in a transmitter, a corresponding descrambler will be required in the receiver. Explain how the data can be recovered. Also, explain modes of transmission where a descrambler (a) must be synchronized to the incoming data and (b) operates without initialization. What type of error, if any, will be introduced in (b)?

8. Consider a channel for which the following sample values of its pulse response are given:

$z_c(0) = 1.0$ $z_c(T) = 0.2$ $z_c(2T) = -0.08$ $z_c(3T) = 0.008$

$z_c(-3T) = 0.001$ $z_c(-2T) = -0.05$ $z_c(-T) = 0.12$

Find (a) the taps coefficient for the three-tap zero forcing equalizer and (b) the output samples for $mT = -2T, -T, 0, T,$ and $2T$.

9. A 20-m Cassegrain antenna, whose receive antenna gain is 54.53 dB at 11.2 GHz and ambient temperature of 275 K, is characterized by the following parameters:

Antenna noise temperature = 60 K.

Waveguide loss = 0.3 dB.

The LNA has a 20-dB gain and an effective noise temperature of 350 K.

If the effective noise temperature of the downconverter is 1500 K, estimate the earth station system noise temperature and the antenna gain referred to the input of the LNA.

10. A satellite repeater system consists of three channels: the uplink, transponder processing unit, and downlink. The uplink has an error probability of 0.1%. The processing unit has an error probability of 0.025%. If the downlink error probability of 1% is desired, calculate:

a. The capacity of each channel.

b. The conditional probability of the whole system.

c. The expected capacity of the overall system.

11. Write a computer program to design an equalizer for specified channel conditions using (a) zero forcing and (b) minimum mean square error criteria.

12. Simulate the probability of bit error for M-ary PSK for variable M and $10^{-5} \le P_e \le 10^{-4}$.

13. Using (3.12), compute the M-PSK system's bit error probability for a range of energy per bit-to-noise power density ratio $(E_b/N_o) = 0, 3, 6,$ and 9 dB.

14. For Problem 13, plot the graphs of P_e using the exact expression and when Q is used. Estimate the difference in percentage between the two results.

15. What can you conclude about coding systems in general as the bit error probability becomes increasingly severe?

References

1. Gargliardi, R.M. (1991). *Satellite communications*. New York: Van Nostrand Reinhold.
2. Tomasi, W. (1987). *Advanced electronic communications systems*. Englewood Cliffs, NJ: Prentice-Hall.
3. Ziemer, R.E., and Tranter, W.H (1988). *Principles of communications: systems, modulation, and noise*. New York: John Wiley.
4. Shannon, C.E. (1948). A mathematical theory of communication. *Bell System Technical Journal*, 27.
5. Rektorys, K. (1969). *Survey of applicable mathematics*. London: Iliffe Books.
6. Holzmann, G.J. (1991). *Design and validation of computer protocols*. Englewood Cliffs, NJ: Prentice-Hall.
7. Held, G (1991). *The complete modem reference*. New York: John Wiley.
8. Qureshi, S. (1982). Adaptive equalization. *IEEE Communications Magazine*, 20.
9. Afshar, A. (1995). *Principles of semiconductor network testing*. London: Butterworth-Heinemann.
10. Kantor, L.Ya. (ed.). (1987). *Handbook of satellite communication and broadcasting*. Boston: Artech House.
11. Kitsuregawa, T. (1990). *Satellite communication antennas: electrical and mechanical design*. Boston: Artech House.

4

Satellite Links

A *satellite link* consists of an uplink (transmit earth station to satellite) and a downlink (satellite to the receive earth station); see Figure 4.1. Signal quality over the uplink depends on how strong the signal is when it leaves the source earth station and how the satellite receives it. Also, on the downlink side, the signal quality depends on how strongly the satellite can retransmit the signal and how the receiving earth station receives the signal.

Satellite link design involves a mathematical approach to the selection of link subsystem variables in such a way that the overall system performance criteria are met. The most important performance criterion is the signal quality, that is, the energy per bit noise density ratio (E_b/N_o) in the information channel, which carries the signal in the form in which it is delivered to the user. As such, in designing a satellite communication system, the designer must attempt to ensure a minimum E_b/N_o in the receiver's baseband channels, which also meets constraints on satellite transmitter power and radio-frequency (RF) bandwidth.

We saw in Chapter 3, Section 3.2 that, for digital transmission, energy per noise density ratio (E_b/N_o) in a baseband channel depends on several factors: carrier-to-noise ratio (C/N) of the receiver, type of modulation used to impress the baseband signal onto the carrier, and the channel's bandwidth. This chapter is concerned mainly with the design and analysis of satellite communication links in terms of the C/N plus interference. As a consequence, a logical step is to calculate (1) the carrier (received) power in an earth station receiver, as well as (2) the noise plus interference power in the receiver, to establish the combined C/N.

4.1 Link Equations

The link equations deal with the calculation of the available C/N over a satellite link. The calculation of the power an earth terminal receives from a satellite transmitter is fundamental to understanding satellite communications and the development of the link equations. A good start is to develop the link equations from the transmission theory perceptive.

Assume that a transmitter T_x in Figure 4.2(a) is equipped with a hypothetical isotropic antenna that radiates spherically in space. At a distance R from

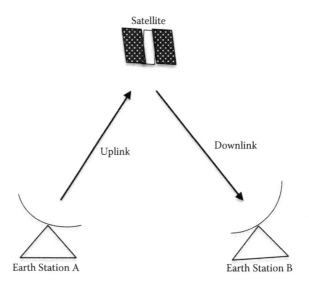

FIGURE 4.1
A simplex link.

the hypothetical isotropic source, the flux density F, crossing the spherical surface, with distance R taken as its radius may be written as

$$F = \frac{P_T}{4\pi R^2} \qquad \text{W/m}^2 \qquad (4.1)$$

where P_T is the total power radiated by the transmitting source.

Because the transmitting antenna would have a gain G_T in the receiver R_x direction, the flux density at the receiver is

$$F = \frac{G_T P_T}{4\pi R^2} \qquad \text{W/m}^2 \qquad (4.2)$$

Chapter 2, Section 2.7 discusses that an antenna is characterized by its effective aperture area. If the effective aperture area is represented by A_e, then the receiving antenna will capture a signal power

$$P_r = FA_e = \frac{G_T P_T}{4\pi R^2} A_e \qquad \text{W} \qquad (4.3)$$

An effective aperture area A_e is a measure of the effective absorption area an antenna presents to an incident plane wave. The effective aperture gain G_r

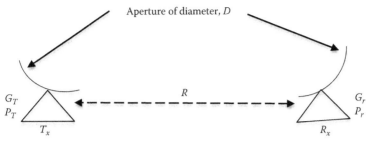

(a) Schematic diagram of a transmission link between
transmitting T_x and receiving R_x antennas.

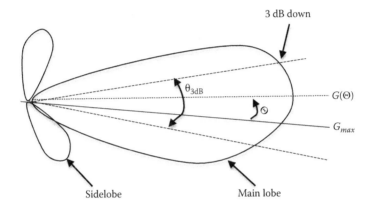

(b) Antenna radiation pattern, where Θ is the depointing angle.

FIGURE 4.2
Basic antenna parameters: (a) schematic diagram of a transmission link between transmitting
T_x and receiving R_x antennas; (b) antenna radiation pattern, where Θ is the depointing angle.

may be expressed by the steradian solid angle of the receive beamwidth Ω to
that of the solid angle of the sphere, specifically

$$G_r = \frac{4\pi}{\Omega} = \frac{4\pi}{\theta_{el}\theta_{az}} \tag{4.4a}$$

At far-field distance, the beamwidth Ω may be represented as the product of
the elevation sector beamwidth, θ_{el}, and azimuth sector beamwidth, θ_{az}. Each
sector's beamwidth may be defined as

$$\theta_{el} = \frac{k_e\lambda}{D_{el}}$$

$$\theta_{az} = \frac{k_e\lambda}{D_{az}} \qquad \text{rad} \tag{4.4b}$$

D_{el} and D_{az} are aperture diameters in elevation and azimuth, respectively. λ and k_e are, respectively, the wavelength and a constant determined by the aperture illumination. The constant k_e is used to emphasize the type of aperture used. It varies between radar and optical bandwidths [1]. The constant k_e is typically measured in radar at the half-power bandwidth point (i.e., 3 dB drop-off in Figure 4.2(b)). For brevity's sake, if we take k_e as unity and $A_e = D_{el}D_{az}$, the receive antenna gain is related to aperture area:

$$A_e = \frac{\lambda^2 G_r}{4\pi} \qquad \text{m}^2 \qquad (4.5a)$$

In satellite communication textbooks, the product of the actual aperture surface area and the efficiency η of the antenna determines the effective aperture area. Hence, Equation (4.5a) is rewritten as

$$A_e = \frac{\lambda^2 G_r}{4\pi\eta} \qquad \text{m}^2 \qquad (4.5b)$$

For unity efficiency, and substituting (4.5) in (4.3), the ratio of the received power to the transmitted power can be written as

$$\frac{P_r}{P_T} = G_r G_T \left(\frac{\lambda}{4\pi R} \right)^2 \qquad (4.6)$$

The power attenuation, a_p, is the value of (4.6) expressed in dB:

$$a_p = 10\log\left(\frac{P_r}{P_T} \right) = G_r + G_T + 20\left[\log\lambda - \log R \right] - 21.984 \qquad (4.7)$$

Note that the gains G_R and G_T are in dB.

The squared component in Equation (4.6) is called the inverse of the *free-space loss*, L_{fs}:

$$L_{fs} = \left(\frac{4\pi R}{\lambda} \right)^2 \qquad (4.8)$$

This expression is frequency dependent: the higher the frequency, the higher the free-space loss. Note that $\lambda = c/f$, where c is the speed of light and f is the propagation frequency. *Free-space loss* is a term traditionally used to calculate the radio link power with the gains of both antennas. The actual relationship by which electromagnetic radiation density diminishes with distance, called *spreading loss*, is just the inverse square relationship, that is, $(1/4\pi R^2)$, and is independent of wavelength.

To have a feel for L_{fs}, a typical geostationary satellite's free-space loss is between 195 and 213 dB for frequencies between 4 and 30 GHz on the assumption of a subsatellite earth station.

The antenna gains expressed above are assumed to have been measured at the antenna electric axis where the gains are maximum ($G = G_{max}$). In practice, there might be some deviation—see Figure 4.2(b)—where the deviation angle from the electric axis is represented by Θ. The gain G can be that of the receive or transmit antenna. Maral and Bousquet [2] have shown that for a small deviation (or depointing) angle, the antenna gain may be written:

$$G(\Theta) = G_{max} - 12\left(\frac{\Theta}{\theta_{3dB}}\right)^2 \qquad \text{dB}$$

(4.9)

The angle θ_{3dB} is the half-power bandwidth of the antenna, which corresponds to the 3 dB drop-off in the antenna gain as a result of the deviation.

From (4.6), the received power P_r can be expressed in terms of other parameters:

$$P_r = P_t G_t G_r \left(\frac{\lambda}{4\pi R}\right)^2 \qquad \text{W}$$

(4.10)

This expression, known as the *Friis transmission equation*, is important in the analysis of any satellite link. The received power P_r is commonly referred to as the *carrier power*, C. For most digital modulation types, including *M*-PSK, the received power is the unmodulated carrier power. The product of the gain G_t and power P_t of the transmitting antenna is called the *effective isotropic radiated power* (EIRP). Alternatively, we write the received power as

$$P_r = EIRP + G_r - L_{fs} \qquad \text{dB}$$

(4.11)

where all the terms in the equation are in dB. In reality, there are other losses, including L_{fs}, that constitute path losses in the link equations. These losses include system loss (due to thermal noise), transmission loss due to ionosphere and precipitation, and directional (pointing) loss. The only loss yet undefined is that due to precipitation (rain), which shall be discussed later in Section 4.1.2. In general, L_{fs} is replaced with the total path loss L_p, that is,

$$P_r = EIRP + G_r - L_p \qquad \text{dB}$$

(4.12)

The C/N can be obtained by dividing the carrier power by the system noise power. The system noise power was defined in Equation (3.8) as

$$N = kTB$$

(4.13)

where k = Boltzmann's constant = 1.38×10^{-23} W/(Hz-K); B = bandwidth required by the desired information rate and modulation scheme, obtainable from Equation (3.6); and T = system noise temperature, which is given by Equation (3.57). It should be noted that the antenna temperature T_a in (3.57) assumes a clear-sky condition. If data are transmitted during rain, rain will increase the antenna noise temperature. In that case, replace T_a in (3.57) with the equivalent antenna temperature.

In view of (4.10) and (4.13), the is written as

$$\frac{C}{N} = \frac{G_r G_T P_T}{kBT} \left(\frac{\lambda}{4\pi R} \right)^2 \tag{4.14}$$

This expression neglects losses other than the free-space loss. As noted earlier in this chapter, a major digital satellite link design condition is to ensure that the E_b/N_o ratio is sufficiently large to guarantee the *bit error rate* (BER) performance criteria are met. The relationship between C/N and E_b/N_o was established in Chapter 3, Equation (3.11) as

$$\frac{E_b}{N_o} = \left(\frac{C}{N} \right) \left(\frac{B}{r_b} \right) \tag{4.15}$$

where r_b is the data bit rate. Thus, substituting the bit rate expression of (3.7) in (4.15):

$$\frac{E_b}{N_o} = \left(\frac{C}{N} \right) \left[\frac{1+\alpha}{\log_2 M} \right] \tag{4.16}$$

where α is the channel filter roll-off factor and M is the possible values or symbols the phase of the carrier takes during a modulation scheme. For an ideal filter, $\alpha = 0$.

The fundamental relationship expressed by (4.14) can be used to calculate the receiver C/N for each link in turn. If we use subscripts u, s, and d, respectively, for the uplink, satellite transponder, and downlink, each link's expressions are developed as follows.

Uplink equations:

$$\left(\frac{C}{N} \right)_u = 10\log(G_T P_T) - 20\log\left(\frac{4\pi R_u}{\lambda_u} \right) + 10\log\left(\frac{G_s}{T_s} \right)$$
$$-10\log(B_s k) + 10\log(L_u) \tag{4.17}$$

where $G_T P_T$ is the EIRP of the transmitting earth station, L_u is other losses associated with the upward transmission, B_s is the transponder bandwidth, and (G_s/T_s) is the *figure of merit* (FOM) of the satellite transponder.

Downlink equations:

$$\left(\frac{C}{N}\right)_d = 10\log\left(G_s P_s\right) - 20\log\left(\frac{4\pi R_d}{\lambda_d}\right) + 10\log\left(\frac{G_d}{T_d}\right)$$
$$-10\log(kB_r) + 10\log(L_d) \tag{4.18}$$

where $G_s P_s$ is the EIRP of the satellite, L_d is other losses associated with the downward transmission, B_r is the receiving earth station's bandwidth, and (G_d/T_d) is the FOM of the receiving earth station.

Having determined each link C/N, we are in a position to establish the combined carrier-to-noise ratio $(C/N)_c$. As seen in Figure 4.1, each single-hop earth station to earth station via a satellite consists of an uplink and a downlink in cascade, which can be modeled as in Figure 4.3. G_s represents the effective transponder gain. C_s and N_s are the carrier and the effective noise powers of the associated links, respectively.

From Figure 4.3, the following expressions can be written:

1. Output carrier power of the satellite transponder:

$$C_d = G_s C_u \tag{4.19}$$

2. Combined carrier power of the single hop:

$$C_c = G_d C_d = G_s G_d C_u \tag{4.20}$$

3. Combined noise power:

$$N_c = N_u G_s G_d + N_d G_d \tag{4.21}$$

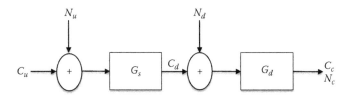

FIGURE 4.3
A model of the combined-link C/N.

In view of (4.19) to (4.21), the combined power ratio in cascade is

$$\frac{N_c}{C_c} = \frac{N_u G_s G_d + N_d G_d}{C_u G_s G_d}$$

$$= \frac{N_u}{C_u} + \frac{N_d}{C_u G_s}$$

$$= \frac{N_u}{C_u} + \frac{N_d}{C_d} \tag{4.22}$$

Thus, the combined C/N is

$$\left(\frac{C}{N}\right)_c = \frac{1}{\left(\frac{C}{N}\right)_u^{-1} + \left(\frac{C}{N}\right)_d^{-1}} \tag{4.23}$$

This expression is equally valid for the combined energy per bit noise density ratio (E_b/N_o).

Example 4.1

Calculate the combined C/N of the system when the input C/N to the satellite transponder is 22.8 dB while the receiving earth station's C/N is 16.4 dB.

SOLUTION

First convert the dB ratio to a power ratio:

$$\left(\frac{C}{N}\right)_u = 10^{2.28}$$
$$\left(\frac{C}{N}\right)_d = 10^{1.64}$$

Substitute these in (4.23) to obtain $\left(\frac{C}{N}\right)_c = 35.5$ or 15.5 dB.

4.1.1 Link Power Budget

Equation (4.12) is often calculated by setting out various terms in tabular form. This tabular presentation of variables in the transmission medium and transmitter/receiver characteristics, such as power gains and losses, is referred to as *link power budget*. An example is given in Table 4.1, noting that −10 dBmW = −40 dBW.

TABLE 4.1

Link Power Budget

Description	Symbols	Positive Factors	Negative Factors
Power required at receiver input	P_r		–116 dB
Receiver losses	L_r	0.45 dB	
Antenna gain (receiver)	G_r		–40 dB
Other losses	L_o	4 dB	
Free-space losses	L_{fs}	206 dB	
Antenna gain (transmitter)	G_T		–42 dB
Transmitter losses	L_T	0.55 dB	
Total		211 dB	–198 dB
Transmitter power		**+13 dB (20 W)**	

4.1.2 Rain Attenuation

Rain attenuation is, perhaps, the most important single phenomenon that impairs transmission of a satellite signal. How the loss due to rain is calculated is the subject of this section. Rain attenuation is modeled as

$$A_r = aLr_r^b \tag{4.24}$$

where:

1. a and b are coefficients that are calculable theoretically from considerations of electromagnetic propagation in spherical raindrops. These coefficients are polarization and frequency dependent based on raindrop characteristics and can be approximated to the following analytical expressions [3]:

$$a = \begin{cases} 4.21x10^{-5}\,f^{2.42}, & f \leq 54GHz \\ 4.09x10^{-2}\,f^{0.699}, & 54 < f \leq 180GHz \end{cases} \tag{4.25a}$$

$$b = \begin{cases} 1.41f^{-0.0779}, & f \leq 25GHz \\ 2.63f^{-0.272}, & 25 < f \leq 164GHz \end{cases} \tag{4.25b}$$

Outside these frequency ranges, the coefficients are equated as zero. If the coefficients are linearly polarized vertically or horizontally, the coefficients for a circularly polarized wave can be calculated using [4]:

$$a_c = 0.5(a_h + a_v) \tag{4.26a}$$

$$b_c = \frac{a_h b_h + a_v b_v}{2a_c} \tag{4.26b}$$

The subscripts of the constants indicate their polarization. For example, subscripts c, h, and v denote circular, horizontal, and vertical polarization, respectively.

2. r_r = rain rate (mm/h). Average values of r_r can be obtained from your country's Department of Meteorology (or its equivalent).

3. L = rain geometric pathlength (km). A simple geometric expression for calculating the pathlength has been given by Ippolito et al. [5] and Lin [6]:

$$L = \frac{L_o}{1 + \dfrac{L_o(r_r - 6.2)}{2636}} \tag{4.27a}$$

where

$$L_o = \frac{H_o - H_g}{\sin\theta} \tag{4.27b}$$

θ = earth station elevation angle (degree), as defined by Equation (2.20); H_g = altitude of earth station (km); and H_o = freezing height (in km), expressed in terms of the earth station's latitude L_{ET}. Thus:

$$H_o = \begin{cases} 4.8, & L_{ET} < 30° \\ 7.8 - 0.1 L_{ET}, & L_{ET} \geq 30° \end{cases} \tag{4.27c}$$

Precipitation (rain) loss (4.24) will be included as part of path losses L_u and L_d in (4.17) and (4.18), respectively, when considering the combined C/N of the communication link.

Figures 4.4 and 4.5 are borne out of Equations (4.24) and (4.27), respectively. Rain effects become severe at wavelengths approaching the water drop size, which is dependent on the type of rainfall. If a satellite link is to be maintained during rainfall, it is imperative that enough extra power be transmitted to overcome the maximum additional attenuation induced by the rain, implying that an accurate assessment of expected loss when evaluating link parameters be made. This attenuation loss is calculated by multiplying the loss per rainstorm pathlength with the mean pathlength: values that can be extracted from Figures 4.4 and 4.5 or, alternatively, using Equations (4.24) and (4.27) after collecting the truth data of the earth station, such as location (latitude, altitude), orientation (elevation angle), and the average rainfall per hour most of the year.

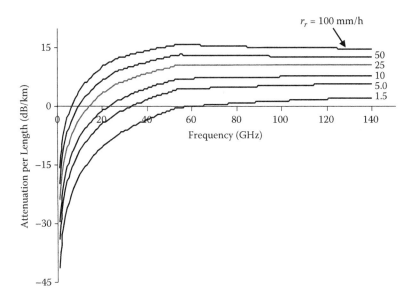

FIGURE 4.4
Rainfall attenuation vs. frequency and rainfall rates, r_r.

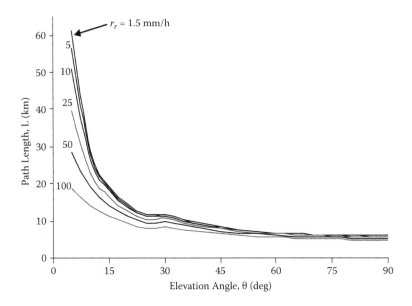

FIGURE 4.5
Average rainstorm pathlength, L, vs. elevation angle for varying rainfall rates, r_r.

Example 4.2

Consider an earth station situated at latitude 5.6°N and an altitude of 100 m. The Bureau of Meteorology gave the average rainfall of 7.85 mm/h, not exceeding 1.09% of the year for the location. If the station is to be oriented to the satellite at 28.5°, estimate the loss due to rain and the availability of the link for 12.5 GHz transmission.

SOLUTION

This example brings us to the concept of outage and availability already discussed in Chapter 2, Section 2.6.5. If $P_o(A_r)$ is the percentage of time (also called percent outage) an attenuation is exceeded, then $\{1 - P_o(A_r)\}$ is the percent availability of the link with an attenuation not exceeding A_r. A_r is the attenuation given by (4.24). Solving this problem:

1. Calculate coefficients a and b using (4.25):

$$a = 0.01918$$

$$b = 1.1523$$

2. Calculate pathlength using (4.27):

$$L = 9.9759 \text{ km}$$

3. Calculate attenuation A_r in dB using (4.24):

$$A_r = 3.13 \text{ dB}$$

4. Availability

The link can withstand and be available $\{1 - (1.09\%)\}$ times of the year if the link design has factored in the effect of rain attenuation. That is, the link is available 98.91% of the year.

4.2 Carrier-to-Noise Plus Interference Ratio

A transmission source cannot be completely ignored as a potential cause of interference simply because its primary transmission is out of the operating band of the satellite links. Interference generated within the transponder can be a result of one or all of the following:

1. Amplification of multiple carriers by the high-power amplifiers, which produces intermodulation products and in turn causes strong interference in other frequency bands.

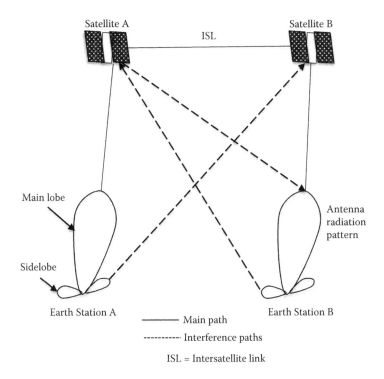

FIGURE 4.6
Adjacent satellite interference.

2. Adjacent channel interference or adjacent transponder interference, which arises in band-limited satellite channels and may cause spectrum spreading.
3. A network of earth stations and satellites operating within proximity of the frequency band, where there is no careful coordination. This effect is explained as follows, using Figure 4.6 as a guide.

Suppose that transmitting earth station (earth station A) radiates a small percentage of its EIRP toward another satellite (satellite B). Often the interference generated by an earth station comes from its antenna sidelobes. This interference signal will appear as noise in the information bandwidth. The noise and signal in a frequency translating transponder will be amplified and reradiated to the receiving earth station, B. As a result, the reradiated interfering uplink and the interfering downlink transmissions combine and appear as noise in the receiver. It is also possible that satellite B would radiate some interfering signal to the receiving earth station, A, which is receiving from satellite A. By regulation, this interference is out-of-band and is usually neglected.

Regulatory agencies (discussed in Chapter 7) provide some guidance on a permissible sidelobe's envelope level relative to a unity isotropic gain

(1 or 0 dB). For example, the U.S. radiocommunications regulator, the Federal Communications Commission (FCC), specifies the following as permissible sidelobe envelope levels [7]:

$$
\begin{array}{ll}
(29 - 25\log\theta)dB & 1^{o} \leq \alpha \leq 7^{o} \\
8dB & 7^{o} \leq \alpha \leq 9.2^{o} \\
(32 - 25\log\theta)dB & 9.2^{o} \leq \alpha \leq 48^{o} \\
-10dB & 48^{o} \leq \alpha \leq 180^{o}
\end{array}
\tag{4.28}
$$

where α is the antenna off-axis (or depointing) angle in degrees. For example, at 6.2°, the sidelobe envelope level must not exceed 9.2 dB above the isotropic (0 dB) level.

The angular separation between satellites, as seen by the earth stations, influences the level of interference generated or received from the sidelobe of the earth station antenna into or from an adjacent satellite. Example 4.3 is given as an illustration of the influence.

Example 4.3

If the angular separation between two geostationary satellites is 3.45°, and a station-keeping accuracy of ±0.5° is assumed, then the worst-case viewing angle is 2.95°, which corresponds to the antenna off-axis, α, of approximately 3°. Hence, using (4.28), the earth station antenna sidelobe's envelope level must not exceed 17.1 dB above the isotropic 0 dB level.

4.2.1 Intersatellite Links

As indicated earlier in Chapter 2, satellite networks often require communications between two satellites via an intersatellite link (ISL), as in Figure 4.6. An ISL is also called a *cross-link*. As a communication link, an ISL has the disadvantage that both transmitter and receiver are spaceborne, thereby limiting operation to both low transmitter power P_t and low figure-of-merit (G/T) values. For long cross-links, it would be necessary to increase the satellite EIRP. This means resorting to transmitting with narrow beams to achieve a higher power concentration. We are constrained by the size of antenna on the satellite, as such narrow beams can be achieved by transmitting at higher frequencies. What are the other trade-offs? The answer will become obvious by modeling cross-linked satellites, as seen in Figure 4.7.

The two satellites (A and B) in Figure 4.7 are assumed to know the exact location of one another, use identical antennas of diameter D and beamwidth ϕ_c pointing at each other, in the same orbit of altitude h, and are not influenced by any aerodynamic drag.

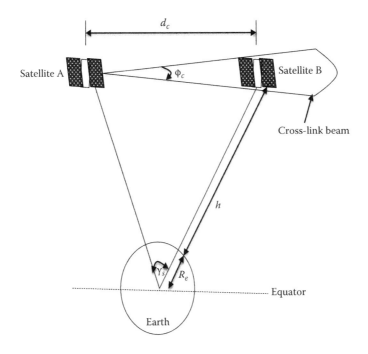

FIGURE 4.7
A model of an intersatellite link.

If the satellites are separated by propagation distance d_c and angle γ_s, and satellite A transmits a signal of power P_t to satellite B of equivalent noise temperature T_e, then, by geometry,

$$d_c = 2(R_e + h)\sin\left(\gamma_s/2\right) \tag{4.29}$$

The propagation distance where maximum line of sight exists is when

$$d_{c(\max)} = 2\sqrt{(R_e + h)^2 - R_e^2} \tag{4.30}$$

For satellites placed at higher altitudes, that is, MEO and above, $h \gg R_e$, so (4.30) approximates to $2h$.

Figure 4.8 plots cross-link distances d_c between two satellites in different orbits as a function of the separation angle. The maximum distances and separation angles, when the two satellites are at different orbits and maintaining line of sight above the earth's atmosphere, are tabulated in Table 4.2.

As seen in Table 4.2, for the two interlinked satellites to maintain line of sight at each of the orbits examined, the latitudinal separation distance between them is more than twice their altitude. Even for a 1° longitudinal separation, the separation distance is in the hundreds of kilometers.

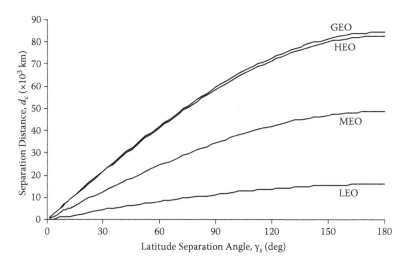

FIGURE 4.8
Satellite separation distance vs. separation angle for various orbits.

We know from (4.4b) that ϕ_c is dependent on antenna diameter and propagation frequency in any sector; that is,

$$\phi_c \approx \frac{\lambda}{D} \tag{4.31}$$

This expression assumes that the aperture illumination constant k_e equals unity. We also know that the receiving antenna gain is related to the aperture area. So, if the antennas are pointing precisely at the desired direction, then from (4.4a) and (4.5a) we can write the transmitting and receiving satellites' antenna gain, respectively, as

$$G_t = \frac{4\pi}{\phi_c^2} \tag{4.32a}$$

$$G_t = 4\pi \frac{A_e}{\lambda^2} \tag{4.32b}$$

TABLE 4.2

Separation Distances and Angles

Orbit	Altitude (km)	d_c (max) (km)	γ_s (°)	At $\gamma_s = 1°$ d_c (km)
LEO	1600	9585.3	73.85	139.24
MEO	18,000	47,057.8	149.67	425.47
HEO	35,000	81,766.9	162.27	722.17
GEO	35,784	83,353.6	162.60	735.86

The illuminated area A_e is a circle circumscribed by satellite B's antenna of diameter D, that is, $A_e = (\pi/4)D^2$. By substituting (4.32) in (4.10), we can write the C/N delivered to the receiving satellite over the ISL as

$$\frac{C}{N} = \frac{\pi P_t D^4}{4N\lambda^2 d_c^2} \tag{4.33}$$

Since, in practice, satellite cross-links are typically in the K-band in the GHz frequency range, we can conveniently express the propagation wavelength λ as $(0.3/f)$, where f is in GHz. Hence,

$$\frac{C}{N} = 8.73 \frac{P_t f^2 D^4}{N d_c^2} \tag{4.34}$$

From this expression, the most practical means of delivering high C/N to the receiving satellite over ISL is by one of the following because in practice we are constrained by the size of the antenna that can be deployed:

1. Decreasing the separation distance d_c
2. Increasing the transmission frequency f
3. Lowering the system front-end noise temperature T_e since N is directly proportional to T_e from Equation (3.8)

The limitation of using an extremely narrow concentration of beamwidth may result in an antenna pointing error due to the inability of satellite A to accurately locate satellite B. Altitudinal error may also occur, that is, the inability of satellite A to accurately orient itself quickly enough to point at the desired direction. These errors are larger than those produced by earth-based tracking antennas. How then do we quantify the uncertainty angle due to inaccurate orientation?

If we denote the altitudinal error by ϕ_{at} and relative location uncertainty by Δr, we can express the total uncertainty angle as

$$\Theta = \phi_{at} + \frac{180\Delta r}{\pi d_c} \text{ deg} \tag{4.35}$$

This expression implies that the transmitting satellite's beamwidth must be wide enough to accommodate these pointing errors. It therefore follows that a trade-off exists, either reducing beamwidth ϕ_c or improving the pointing accuracy.

Despite attempts to stabilize the satellites in their orbit, it is difficult to envision precision pointing every time. It is therefore necessary to consider a situation where uncertainty can be visualized. If each satellite antenna has pointing error Θ, each antenna's gain will be a function of $G(\Theta)$, as in (4.9). Let us denote the transmitting and receiving antenna gains by $G_t(\Theta_t)$ and $G_r(\Theta_r)$, respectively, and their pointing errors as Θ_t and Θ_r, respectively. We can rewrite C/N delivered at the receiving satellite, B, as

$$\frac{C}{N} = \frac{P_t G_t(\Theta_t) G_r(\Theta_r) \lambda^2}{(4\pi)^2 N d_c^2} \tag{4.36}$$

Upon application of (4.9) in (4.36), the carrier power delivered at the receiving end can be estimated. Note that the half-power beamwidth θ_{3dB} in (4.9) equals ϕ_c in (4.31).

As noted earlier in this chapter, a major digital satellite link design condition is to ensure that the E_b/N_o ratio is sufficiently large to guarantee that the *bit error rate* (BER) performance criteria are met. From this established functional relationship between C/N and E_b/N_o, given by Equation (3.11), and for an *M*-PSK system, the decoding E_b/N_o can be written as

$$\frac{E_b}{N_o} = \frac{P_t G_t(\Theta_t) G_r(\Theta_r) \lambda^2}{(4\pi)^2 \kappa T_e d_c^2 r_b} \tag{4.37}$$

where κ is the Boltzmann constant ($= 1.38 \times 10^{-23}$ W/Hz-K).

Figure 4.9 shows the variation in the data transfer rate with increasing antenna size for two LEOs maintaining line of sight over the earth's

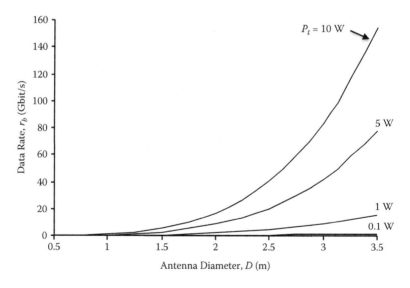

FIGURE 4.9
Cross-link transmission data rate vs. antenna diameter. Parameters used: $f = 14.5$ GHz, $\eta = 65\%$, $d_c = 9{,}598$ km, $E_b/N_o = 9.6$ dB, $T_e = 1000$ K.

TABLE 4.3

Data Rate for Low Power

D (m)	r_b (Mbit/s)
0.5	0.645
1.0	10.313
1.5	52.212
2.0	156.015
2.5	402.869
3.0	835.390
3.5	1547.663

atmosphere, with several transmitter powers, P_t. The antennas are assumed to be pointing precisely to the desired direction. To have a feel for the data rate that can be transmitted across the cross-link at 14.5 GHz frequency with an acceptable probability of error ($P_e = 10^{-5}$) for low transmitter power ($P_t = 0.1$ W), the computed data rate is tabulated in Table 4.3.

Figure 4.10 demonstrates that increasing the antenna diameter size produces increasing cross-link E_b/N_o until corresponding narrowing beamwidths begin to cause a rapid antenna gain falloff for a particular pointing error. For the graphs in Figure 4.10, the transmitting and receiving antennas are assumed identical, and the pointing error at each end of the antennas is assumed to be the same; that is, $\Theta_t = \Theta_r = \Theta$.

To ensure that pointing error is maintained to the barest minimum, a form of automatic tracking (simply autotracking) system should be installed at the receiver. The function of the autotracking system is to track the arrival of the

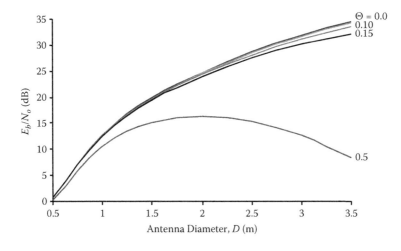

FIGURE 4.10
Cross-link E_b/N_o vs. antenna diameter. Parameters used: $f = 14.5$ GHz, $\eta = 65\%$, $r_b = 50$ Mbit/s, $d_c = 9598$ km, $T_e = 1000$ K.

transmitter beam and point it back at the receiver antenna. Chapter 3, Section 3.4.2 discusses types of antenna tracking as well as analytical description. The curves shown by Figure 4.10 are instrumental in determining the trade-off between designing and installing an automatic tracking system or simply allowing an open-loop mispointing.

4.2.2 Modeling Interference Plus Noise

For brevity, the interference is modeled as additive interfering signals on the uplink with respective powers, say, $I_{u1}, I_{u2}, ..., I_{uk}$, within the bandwidth of the desired carrier. Therefore, the total uplink interference plus noise power can be written as

$$N_{IU} = N_u + \sum_{i=1}^{k} I_{ui} \tag{4.38}$$

The uplink C/N plus interference ratio can be written as

$$\left(\frac{C}{N}\right)_{UI} = \left(\frac{C}{N}\right)_u + \sum_{i=1}^{k} \frac{C_u}{I_{ui}}$$

$$= \left(\frac{C}{N}\right)_u + \left(\frac{C}{I}\right)_u \tag{4.39}$$

Follow the uplink procedure for the downlink case and consider additive interfering signals on the downlink with respective powers, say, $I_{d1}, I_{d2}, ..., I_{dk}$, within the bandwidth of the desired carrier. As such, the corresponding C/N plus interference ratio can be written as

$$\left(\frac{C}{N}\right)_{DI} = \left(\frac{C}{N}\right)_d + \sum_{i=1}^{k} \frac{C_d}{I_{di}}$$

$$= \left(\frac{C}{N}\right)_d + \left(\frac{C}{I}\right)_d \tag{4.40}$$

Thus, the combined C/N for a single-hop transmission is

$$\frac{C}{N} = \frac{1}{\left(C/N\right)_{UI}^{-1} + \left(C/N\right)_{DI}^{-1}} \tag{4.41}$$

This expression is the most widely used equation in satellite system engineering. Care must be taken when applying (4.41). For example, when rain occurs on the downlink only, the C/N and interference are attenuated equally. As such, the downlink carrier-to-interference ratio remains at the clear-sky value and Equation (4.40) is used, meaning that the rain attenuation factor is not included. However, if rain occurs in both uplink and downlink, the effect of rain-induced attenuation must be included in the link calculation, which is to further reduce the C/N plus interference ratio of the link being considered.

The overall system quality E_b/N_o, the bit energy-to-noise plus interference density ratio, can be computed from (4.41) and (3.11).

Example 4.4

A satellite system uplink was designed so that the transponder C/N plus interference ratio is 22.5 dB. The link is cascaded with the downlink for which the receiver C/N plus interference ratio is 16.25 dB. Estimate the overall C/N of the cascaded link.

SOLUTION

Convert the C/N plus interference ratios into power ratio form:

$$\left(\frac{C}{N}\right)_{UI} = 10^{2.25} = 177.828$$

and

$$\left(\frac{C}{N}\right)_{DI} = 10^{1.625} = 42.1697$$

Then using (4.41), the overall C/N at the cascaded link output becomes

$$\frac{C}{N} = \frac{1}{(177.828)^{-1} + (42.1697)^{-1}} = 34.087 \text{ or } 15.33 \text{ dB}$$

4.3 Summary

The quality of signals received by the satellite transponder and that retransmitted and received by the receiving earth station is important if successful information transfer via the satellite is to be achieved. Within constraints of transmitter power and information channel bandwidth, a communication system must be designed to meet certain minimum performance standards. The most important performance standard is the bit energy per noise density ratio in the information channel, which carries the signals in a format in which they are delivered to the end users. This chapter explains the process of designing and calculating the C/N as a measure of the system performance standard.

Problems

1. A commercial satellite communication system is to be deployed in an orbit and is noted to use the 6/4 GHz band. The system has the following parameters:

   Satellite:

   Frequencies:

   14.75 GHz up

   11.36 GHz down

   Antennas: 1.425° beamwidth up and down. (Note that this is 3 dB measurement.)

   Transponder:

   Bandwidth = 50 MHz

   Saturated output power = 20.02 W

   Input noise temperature = 226.5°C

   Gain (up to point of saturation) = 119.2 dB

   Path losses in benign environment:

   201 dB at 14.75 GHz

   195 dB at 11.36 GHz

   Earth station:

   Antenna: Diameter 3 m

   Aperture efficiencies:

   64.5% at 1.47 GHz

   61.25% at 11.36 GHz

   Ohmic loss between the antenna and LNA = 1.0 dB

   Receiver (IF) bandwidth = 50 MHz

   Calculate the transmitter power required to just saturate the transponder output amplifier for an earth station at the edge of the coverage zone of the satellite antenna. Ignoring any noise contributions from the satellite transponder, estimate (a) the earth station's G/T required to give a C/N of 17.92 dB in the earth station IF amplifier and (b) the earth station noise temperature allowing for a loss figure of 0.5 dB.

Find the C/N in the transponder when its output is just saturated. Also find the C/N in the earth receiver IF amplifier when the transponder noise is included.

2. A satellite carrying a 9.8-GHz continuous-wave beacon transmitter is located in geosynchronous orbit 37,586 km from an earth station. The beacon's output power is 0.3 W and feeds an antenna of 19 dB gain toward the earth station. The antenna is 3.65 m in diameter with an aperture efficiency of 62.5%.

   a. Calculate the satellite EIRP.

   b. Calculate the receiving antenna gain.

   c. Calculate the path loss.

   d. Calculate the received power.

   If the overall system noise of the earth station is 1189K, calculate:

   e. The earth station G/T.

   f. The received noise power in a 115-Hz noise bandwidth.

   g. The receiver C/N in the above noise bandwidth.

3. A 14.5-GHz transponder receiver has a G/T of 3.4 dB/K and a saturation flux density of −75.83 dBW/m². The transponder's bandwidth is 73 MHz, which is centered at 14.615 GHz up and 11.455 GHz down. The satellite is used to set up a 73-MHz bandwidth link between two earth stations. The 11-GHz transmitter part of the transponder has a saturated EIRP of 43.89 dBW. The uplink and downlink pathlengths are 41,257 km. Both the transmitting and receiving antennas use 3.85 m diameter dishes and aperture efficiencies of 62.3%. The receiving earth station is designed to have an overall noise temperature of 120.8 K. Calculate:

   a. The uplink EIRP, transmitter output power, and C/N.

   b. The downlink C/N.

   c. The overall system C/N.

4. A satellite transmits with an EIRP of 46 dBW. Calculate the received C/N if the bandwidth is 35 MHz and the receiver has a G/T of 25 dB/K. Assume the distance between the earth and the satellite is 35,786 km.

5. Your company has won a special contract. The contract involves the design and commissioning of an earth station, able to meet relevant international telephony and broadcasting standards. The customer specifies the following as the essential features of the earth station:

   A usable bandwidth of 30 GHz.

   Capable of receiving and transmitting 99.55% of the 3.2 Gbit/s data at all times.

When receiving, it must be sensitive to a carrier power of at least −120 dBW. It is envisaged that the earth station will be situated approximately at longitude 132°E, latitude 10°S of the equator.

Your design should allow for future expansion.

6. For a link to be maintained between a geosatellite and an earth station of about 35,786 km apart, calculate the carrier power (in dB) necessary given the following antenna characteristics:

   60% efficiency

   2.5 cm wavelength

   1.3 m diameter transmitting dish

   1.0 m diameter receiving dish

   To ensure high service quality, it is suggested that the carrier power at the receiver input be at least −116 dBW.

7. A satellite transmits with an EIRP of 46 dBW. Calculate the received C/N if the bandwidth is 35 MHz and the receiver has a G/T of 25 dB/K. Assume the distance between the earth and the satellite is 35,786 km.

8. A broadcast satellite is located in Sydney with the uplinks centered at 20 GHz and the downlinks at 12 GHz. Calculate the uplink rain attenuation at Brisbane and downlink attenuation at Melbourne that will not exceed 99% of the time.

9. A LEO satellite, 1250 km from earth, with a transmitter power of 15 W illuminates a terrestrial circular zone with an approximate radius of 575 km. Calculate the power flux density on the ground in dBW/m^2 (a) for the LEO, (b) if the satellite is HEO, and (c) if the same satellite is positioned in geostationary orbit. (d) What advice will you provide a customer that is considering positioning a satellite in one of these orbits?

10. A satellite transmits with an EIRP of 46 dBW. Calculate the received C/N if the bandwidth is 35 MHz and the receiver has a G/T of 25 dB/K. Assume the distance between the earth and the satellite is 35,000 km.

References

1. Kolawole, M.O. (2003). *Radar systems, peak detection and tracking.* Oxford: Elsevier Science.
2. Maral, G., and Bousquet, M. (1993). *Satellite communications systems.* New York: John Wiley.

3. Gagliardi, R.M (1991). *Satellite communications.* New York: Van Nostrand Reinhold.

4. Tri, T.H. (1986). *Digital satellite communications.* New York: McGraw-Hill.

5. Ippolito, L.J., Kaul, R.D., and Wallace, R.G. (1986). *Propagation effects handbook for satellite systems design.* NASA reference publication 1082(03).

6. Lin, S.H. (1979). Empirical rain attenuation model for earth-satellite path. *IEEE Transactions on Communications,* 27(5), 812–817.

7. Pritchard, W.L., and Sciulli, J.A. (1986). *Satellite communication systems engineering.* Englewood Cliffs, NJ: Prentice-Hall.

5

Communication Networks and Systems

Communication involves the transfer of information between a source and a user. Many sources may attempt to transfer information through one satellite transponder with the result that the available communication capacity of the transponder is shared between several earth terminals. This sharing technique is called *multiple access* (MA). The term *access*, in satellite transmission systems, usually refers to multiple access. Sharing can be in many formats, such as sharing the transponder bandwidth in separate frequency slots (frequency division multiple access [FDMA]), sharing the transponder availability in time slots (time division multiple access [TDMA]), or allowing coded signals to overlap in time and frequency (code division multiple access [CDMA]). D in the acronyms stands for "division." Division implies switching. The various transmission switching techniques are discussed in much detail in this chapter.

5.1 Principles of Multiple Access

As an illustration, consider a satellite transponder of B bandwidth with n channels as shown in Figure 5.1. When the transponder is in an operational state at any time t_i, each channel amplifies every carrier f_i of the earth terminals ES_i (where $i = 1, 2, ..., n$). If no safeguards are in place, it is likely that several carriers will occupy a particular channel simultaneously and will mutually interfere.

To avoid such mutual interference, it is important that the receivers of other earth terminals be able to discriminate among the received carriers. Discrimination can be achieved by any of three techniques: filtering, temporal gating, or signature assignment [1].

1. Filtering. Take, for instance, the carrier energies to be in the frequency domain. If we subdivide the available channel bandwidth into a number, say, n, as shown in Figure 5.1, and assign a subchannel to each user upon that user's request, then the receiver can discriminate among carriers by filtering. This multiple-access

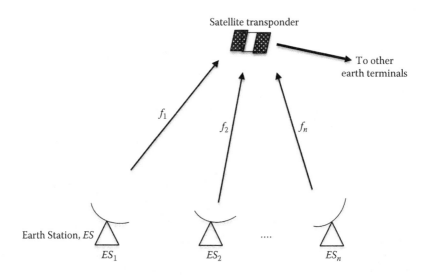

FIGURE 5.1
Principle of multiple access.

method is the principle of FDMA. It is commonly used in wire-line channels to accommodate multiple users for voice and data transmission.

2. Temporal gating. If, for instance, we subdivide the frame duration T_s into, say, n nonoverlapping subintervals, each of duration T_s/n, and assign each user to a particular time slot within each frame, then several carrier energies can be received sequentially by the receiver. The receiver can thus discriminate among carriers by temporal gating. This multiple-access technique is the principle of TDMA. It is frequently used in data and digital voice transmission.

It can be observed from the preceding definitions of FDMA and TDMA that the channel is basically partitioned into independent single-user channels, that is, nonoverlapping frequency bands or time slots, respectively. FDMA and TDMA techniques tend to be insufficient in an environment where [2]:

1. The transmission from various users is *bursty*; that is, information transmission from a single user is separated by periods of silence, simply no transmission. An example of this case is in digital mobile cellular communication systems where digitized voice signals typically contain long pauses.

2. Transmission from several users has a low duty cycle; that is, a situation where the periods of no transmission are greater than the periods of transmission.

An alternative to FDMA and TDMA is to allow more than one user to share a channel or subchannel by use of *signature assignment*.

3. Signature assignment. This is a process by which a user is assigned a unique code (or signature) sequence that allows the user to modulate and spread the information-bearing signal across the assigned frequency band. As a result, signals from various users can be separated at the receiver by cross-correlation* of the received signal with each of the possible user signatures. This process ensures identification of the carriers even when all of the carriers simultaneously occupy the same frequency band. If the code sequences are designed to have relatively small cross-correlations, then the effect of cross talk inherent in the demodulation of the signals will be minimized. This multiple-access technique is the principle of CDMA. Signature assignment is most often realized by means of pseudorandom codes (or pseudonoise codes)—hence the name code division multiple access (CDMA). The effect of using codes is broadening the carrier spectrum in comparison with other schemes that may have been modulated by the carrier information alone. This is why CDMA is also called *spread-spectrum multiple access* (SSMA). Spreading of the carrier spectrum has an inherent antijam advantage. For this reason, the designation SSMA is usually used in conjunction with military systems, whereas CDMA is generally reserved for commercial usage. The major telecommunication provider in Australia, Telstra, has utilized the CDMA techniques to provide telecommunication services in rural areas.

An alternative method to CDMA is *nonspread random access* (NSRA). In this method, when two users attempt to use a common channel simultaneously,

---

* For completeness, we shed some light on what correlation is about. Correlation is a measure of the similarity or relatedness between waveforms. For instance, consider two waveforms $x_1(t)$ and $x_2(t)$. These waveforms may not necessarily be periodic or confined to a finite time interval. Then the *correlation* between them (or more precisely the *average cross-correlation*), between $x_1(t)$ and $x_2(t)$, is defined as

$$R_{12}(\tau) = \lim_{T \to \infty} \frac{1}{T} \int_{-T/2}^{T/2} x_1(t) x_2(t + \tau) dt$$

If $x_1(t)$ and $x_2(t)$ are periodic with the same fundamental period, $T_o = 1/f_o$, then

$$R_{12}(\tau) = \frac{1}{T_o} \int_{-T_o/2}^{T_o/2} x_1(t) x_2(t + \tau) dt$$

If $x_1(t)$ and $x_2(t)$ are nonperiodic pulse-type waveforms, then

$$R_{12}(\tau) = \int_{-\infty}^{\infty} x_1(t) x_2(t + \tau) dt$$

The value of $R_{12}(\tau)$ can be negative, positive, or zero. Where values are registered, their strength determines the correlation type (i.e., either weak or strong). If $R_{12}(\tau) = 0$, then it is *uncorrelated* or *noncoherent*.

their transmissions collide and interfere with each other, resulting in information loss and possibly requiring retransmission. To avoid retransmission, protocols are developed for rescheduling transmissions in nonspread cases. A protocol is a set of necessary procedures or a format that establishes the timing, instructs the processors, and recognizes the messages (more is said about multiple-access protocols in Section 5.3).

In essence, depending on the multiplexing and modulation techniques employed, a range of multiple-access combinations can be derived from these three basic techniques. The hybrid-access types, for example, could be [3, 4]:

1. Frequency/time/code division (FD/TD/CDMA)
2. Frequency/time division (FD/TDMA)
3. Time/code division (TD/CDMA)
4. Frequency/code division (FD/CDMA)

It must be recognized that each carrier can employ any modulation scheme it prefers, for instance, switching or keying the amplitude (amplitude shift keying [ASK]), frequency (frequency shift keying [FSK]), or phase (phase shift keying [PSK]) of the carrier in accordance with the information binary digits. To describe a multiple-access system completely, it is necessary to indicate details of the multiplexing (for instance, frequency division multiplexer [FDM]) and the modulation method (frequency modulation [FM]) for the channel group or carrier to be transmitted. A typical designation employed by the INTELSAT network is FDM-FM-FDMA. FDM-FM-FDMA utilizes a multichannel-per-carrier transmission technique. For instance, the transmitting earth station frequency division multiplexes several single-sideband suppressed carrier telephone channels into one carrier baseband assembly, which frequency modulates a radiofrequency (RF) carrier and in turn transmits it to an FDMA satellite transponder.

5.1.1 Frequency Division Multiple Access (FDMA)

The FDMA scheme allows the partitioning of a bandwidth-limited communication channel into a set of independent lower-speed channels, each of which utilizes its permanently assigned portion of the total frequency spectrum [5]. Each frequency slot contains a unique pair of frequencies needed for sending its digital signals.

A basic FDMA system is shown in Figure 5.2, where each uplink earth station, ES_i, in the network transmits one or more carriers at different carrier frequencies, f_i, to the satellite transponder. Each carrier is assigned a frequency band with a small *guard band* (or safety zone) to avoid electrical overlapping of signals between adjacent carriers (see Figure 5.2(b)). The number of subdivisions permissible, which in turn determines the transmission

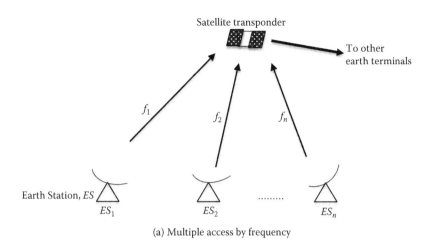

(a) Multiple access by frequency

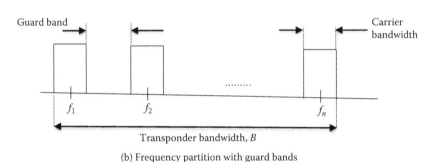

(b) Frequency partition with guard bands

FIGURE 5.2
Principle of FDMA system: (a) multiple access by frequency; (b) frequency partition with guard bands.

capacity, depends on the satellite transponder's channels and bandwidth (this dependence will become obvious by Equation (5.7)).

The FDMA scheme has some advantages and disadvantages. A major limitation of an FDMA system arises from the need for guard bands between adjacent channels in order to avert interference from adjacent channels. These guard bands impose a practical limit on the efficiency of an FDMA system. A secondary disadvantage is the need to control the transmitting power of earth stations in such a way that the carrier powers at the satellite input are the same in order to avoid the capture effect. Despite these disadvantages, FDMA is the oldest technique and will remain the most widely used because of investments already made in it. Major advantages of FDMA are its simplicity and relatively low cost in applications, particularly in multiplexing unclustered terminal groups whose aggregate bit rate limit is not constrained.

5.1.2 Single Channel per Carrier

Among the various transmission schemes corresponding to different combinations of multiplexing and modulation is the *single channel per carrier* (SCPC) scheme. Traffic routing by this scheme is performed according to the one-carrier-per-link principle [6]. For example, each voice (telephone) channel is independently modulated by a separate carrier and is transmitted to the satellite transponder on an FDMA basis. A 36-MHz transponder can carry as many as 800 or more voice channels.

The SCPC/FDMA scheme offers technical and economic advantages, particularly in the case of low-traffic flow between two or more earth stations. However, optimum utilization of the channel transmission capacity is achieved only when channels are pooled and free carriers are assigned on demand to individual stations. Such a system became known under the designation SPADE (single channel per carrier, pulse code modulation (PCM) multiple access, demand assignment equipment).

5.1.3 Time Division Multiple Access (TDMA)

TDMA is characterized by access to the channel during a time slot. Figure 5.3 shows the operation of a network according to the principle of TDMA. As an illustration, the uplink earth stations, ES_i, transmit bursts I_i occupying all of the channel bandwidth during a specified time T_{bi}. A burst corresponds to the transfer of traffic from each uplink station. A burst transmission is inserted within a longer time structure of duration T_s called a *frame period*—a periodic structure within which all uplink stations must transmit. A frame is formed at satellite level. A typical structure of a frame is shown in Figure 5.4.

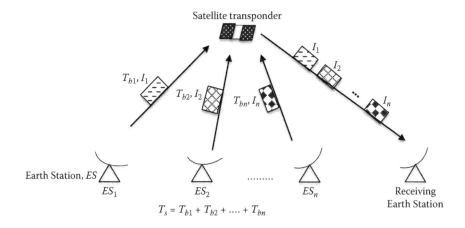

FIGURE 5.3
Operational concept of the TDMA system. Each station ES_i transmits data I_i at assigned time T_{bi} within frame time T_s at the same frequency, where $i = 1, 2, ..., n$.

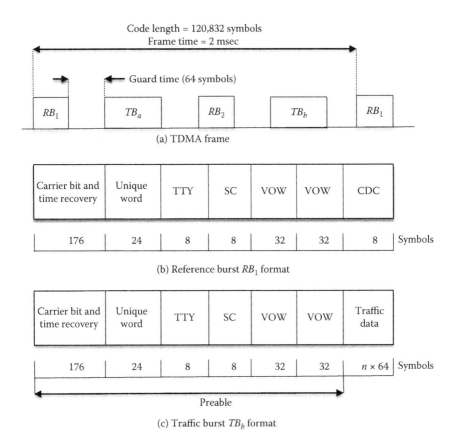

FIGURE 5.4
Typical TDMA structure by INTELSAT/EUTELSAT standard: (a) TDMA frame; (b) reference burst RB_1 format; (c) traffic data TB_b format.

A frame consists of all the bursts (RB_i, TB_i) by the uplink earth stations placed one after another, providing that transmission synchronization of the earth stations is correct.

To correct for synchronization imperfections, a period without transmission, called a *guard time*, similar to guard band in FDMA, is provided between each burst. CDC represents guard time in Figure 5.4(b and c). The structure of a burst (e.g., RB_i, TB_i) is illustrated in Figure 5.4(b and c). The structure consists of a header (or preamble) and a traffic field or the control and delay field (CDC). The traffic field is located at the end of the header and corresponds to the transmission of useful information.

In the case of the one-carrier-per-station method, where the burst transmitted by a station carries all the information from the station to other stations, the traffic field is structured in *subbursts*, which correspond to the information transmitted by the uplink station to each of the other stations. The field

SC is used as service channel, which contains alarms and various network management information.

The header has many functions, which permit [7]:

Demodulation of the receiving earth station to recover the carrier generated by the local oscillator at the uplink station. This is particularly so for the case of coherent demodulation. The header in this regard provides a constant carrier phase.

The receiving (downlink) earth station to identify the start of a burst by detecting a "unique word," that is, a group of bits. This unique word enables the receiver to resolve carrier phase ambiguity (in the case of coherent demodulation) and allows a receive station's detector to synchronize its clock with the symbol rate. In this regard, the header would provide alternating opposite phases.

The transfer of service messages between stations (telephone and telex/facsimile) and signaling via TTY and VOW fields.

If l users are multiplexed to the transmitting station ES_1, as in Figure 5.5, then the bit rate r_b of the multiplexer would be

$$r_b = \sum_{i=1}^{l} r_{bi} \tag{5.1}$$

As a consequence, the transmission rate of transmitting station ES_1 will be

$$R_r = r_b \frac{T_s}{T_b} \text{ bit/s} \tag{5.2}$$

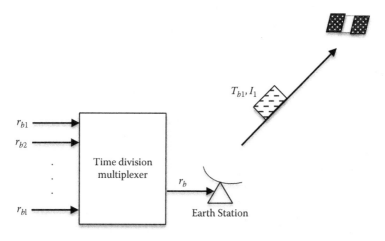

FIGURE 5.5
TDMA: Bit rate r_b for one user.

The throughput of TDMA depends on the number of bursts (number of accesses or traffic stations), n_p, to the frame. If we allow x to be the number of bits in the header and y the number of bits in the guard time, and assume that the frame contains m reference bursts, then the throughput can be written as

$$\eta = 1 - \frac{(n_p + m)(x + y)}{R_r T_s} \tag{5.3}$$

Example 5.1

Consider the frame format of the INTERSAT/EUTELSAT given by Figure 5.4. If an offset-quadrature phase shift keying (OQPSK) modulation scheme is used for transmission, express and plot the TDMA throughput for 200 accesses.

SOLUTION

We can write the following:

$$R_r = 0.1208 \text{ Gbit/s, since 1 symbol} = 2 \text{ bits and } T_s = 2T_b$$

$$T_s = 2 \text{ ms}$$

$$m = 2$$

$$x = 560$$

$$y = 128$$

Substituting these values in Equation (5.3), we can write the throughput as a linear function of the number of traffic stations or accesses n_p:

$$\eta = 1 - 0.00285(n_p + 2) \tag{5.4}$$

Figure 5.6 is a plot of Equation (5.4) and shows a gradual decrease in throughput with the number of users continuously accessing the satellite. The unity throughput ($\eta \approx 1$) corresponds to a single carrier passing through the satellite transponder.

TDMA has certain advantages:

1. All stations transmit and receive on the same frequency regardless of the origin of the burst. This simplifies tuning.
2. At any instant only a single carrier occupies all the channel bandwidth. This ensures that there are no intermodulation products.
3. Transmission throughput is high even for a large number of users. For instance, from Figure 5.6, $\eta = 0.6523$ (65.23%) for $l = 120$.
4. Control is achieved at the satellite level and requires no control of the transmitting power of the uplink stations.

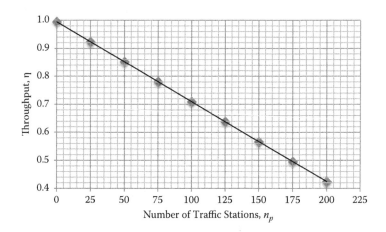

FIGURE 5.6
TDMA throughput for an INTELSAT/EUTELSAT frame.

However, the need for synchronization and the need to dimension the station for transmission at high throughput can be considered the main disadvantages of TDMA systems.

5.1.4 Code Division Multiple Access (CDMA)

CDMA operates on the principle of spread-spectrum transmission; see Figure 5.7. The uniquely separable address *signatures* (codes) are embedded within the uplink carrier waveform. Each uplink uses the entire satellite bandwidth and transmits through the satellite whenever needed, with all active stations superimposing their waveforms on the downlink. As such, no time or frequency separation is required. Carrier separation is achieved at each earth station by identifying the carrier with a proper signature.

The uplink station, as in Figure 5.7(b), spreads the user's spectrum of Figure 5.7(a). The spread spectrum might contain some noise and other interference when it is retransmitted. The receiver recovers the useful information by reducing the spectrum of the carrier transmitted in its original bandwidth, as in Figure 5.7(c). The effect of noise and other interference has been suppressed in Figure 5.7(c) for brevity. This operation simultaneously spreads the spectrum of other users in such a way that they appear as noise of low spectral density.

It should be noted that one could not simply use codes of arbitrary different phases to provide CDMA because the codes have high-autocorrelation sidelobes at the subsequent periods. Furthermore, the power spectral density of the codes has line components at frequencies corresponding to each of the code periods [8].

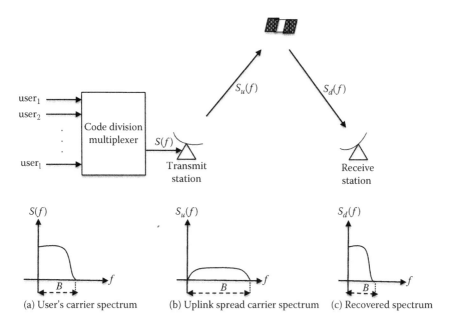

(a) User's carrier spectrum   (b) Uplink spread carrier spectrum   (c) Recovered spectrum

FIGURE 5.7
CDMA system: (a) user's carrier spectrum; (b) uplink spread carrier spectrum; (c) recovered spectrum.

The spreading ratio is determined primarily by the code ratio κ_c/r_c and can be achieved with either low-rate channel codes or long address codes, where

$$\kappa_c = T_s B \tag{5.5}$$

$$r_c = \frac{1}{T_s r_b} \tag{5.6}$$

Thus, the spreading ratio $\kappa_c/r_c = B/r_b$. This ratio is commonly referred to as the spreading ratio of the code modulation or CDMA bandwidth expansion factor. In some texts, this ratio is halved because the carrier bandwidth is taken as $B/2$.

CDMA has some advantages:

1. It is simple to operate.
2. It does not require any transmission synchronization between stations. The only synchronization required is that of the receiver to the sequence of the received carrier.

3. It offers sufficient protection against interference from other stations and that due to multiple paths. This makes CDMA attractive for networks of small stations with large antenna beamwidths and for satellite communication with mobiles.

However, on average, its main disadvantage is the low throughput demonstrated in the next section.

5.2 Capacity Comparison of Multiple-Access Methods

The capacity of multiple-access (FDMA, TDMA, and CDMA) methods can be examined in terms of the information rate that each method achieves in an ideal *additive white Gaussian noise* (AWGN) channel of finite bandwidth B [9, 10]. For simplicity, we assume a uniform number of users, say, l. Each user has a similar average carrier power $C_i = C$, for all values of i users where $i = 1, 2, ..., l$.

In FDMA, each user is allocated a bandwidth (B/l). Hence, in view of Equations (3.10) and (3.19), the capacity for each user can be expressed as

$$C_{FD} = \frac{B}{l} \log_2 \left(1 + \frac{C}{\left(B/l \right) N_o} \right) \tag{5.7}$$

Superficially, this expression indicates that for a fixed bandwidth B, the total capacity goes to infinity as the number of users increases linearly with l. This is not true, as will be demonstrated later. By writing the carrier power C in terms of energy of a single bit E_b as in (3.11), the normalized total capacity C_n for the FDMA system can be expressed as

$$C_n = \log_2 \left(1 + C_n \frac{E_b}{N_o} \right) \tag{5.8}$$

where $C_n = l r_b / B$ is the total bit rate for all l users per unit bandwidth, and r_b is the input digital bit rate of the user (bit/s). Alternatively,

$$\frac{E_b}{N_o} = \frac{2^{C_n} - 1}{C_n} \tag{5.9}$$

Figure 5.8 shows the graph of C_n vs. E_b/N_o. We observe that C_n increases as E_b/N_o increases above the minimum value of $\log_e 2$, which is the absolute

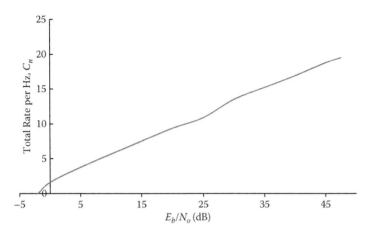

FIGURE 5.8
FDMA total capacity response.

minimum value, demonstrated by Equation (3.22), imposed on a finite channel capacity for an error-free transmission.

In the TDMA system, each user is allowed to transmit for T_s/l of the time through the channel of bandwidth B. The total capacity per user can be expressed as

$$C_{TD} = B\frac{T_s}{l}\log_2\left(1 + \frac{C}{B\left(\frac{T_s}{l}\right)N_o}\right)$$

(5.10)

This expression is identical to that of the FDMA system in (5.7). In the TDMA system, however, it may not be possible for the transmitter to sustain an average carrier power for large l in practice. Hence, there is a practical limit beyond which the transmitter power cannot be increased as the number of users increases.

The evaluation of system capacity in the case of CDMA depends on the level of cooperation among the users. Two extreme cases (*noncooperative* CDMA and *cooperative* CDMA) are considered. In both cases, each uplink user transmits a pseudorandom code or signal and uses the entire satellite bandwidth, B. Each signal distribution is Gaussian.

5.2.1 Case I: Noncooperative CDMA

In the noncooperative case, the receiver for each user code or signal does not know the spreading waveforms of the other users or chooses to ignore the spreading waveforms in the demodulation process. As a consequence, the other user's codes or signals will appear as interference at the receiver of each user.

For developmental purposes, we consider the system receiver to comprise a bank of l single-user receivers. Each user's pseudorandom code is corrupted by Gaussian interference and additive noise with corresponding powers $(l-1)C$ and N $(= BN_o)$. Following Equation (3.19), we can write the capacity per user as

$$C_l = B \log_2 \left(1 + \frac{C}{C(l-1)+N} \right)$$

(5.11)

Rearranging (5.11) in terms of energy of a single bit per user per bandwidth, we have

$$\frac{C}{B} = \log_2 \left[1 + \frac{C/N}{B \left\{ 1 + \frac{C}{N}(l-1) \right\}} \right]$$

$$= \log_2 \left[1 + \frac{r_b/B}{B} \left(\frac{E_b/N_o}{1 + r_b/B(l-1)E_b/N_o} \right) \right]$$

(5.12)

which can be rewritten as

$$C_x = \log_2 \left[1 + C_x \left\{ \frac{E_b/N_o}{1 + C_x(l-1)E_b/N_o} \right\} \right]$$

(5.13)

Or, equivalently,

$$\frac{E_b}{N_o} = \frac{2^{C_x} - 1}{C_x[l - 2^{C_x}(l-1)]}$$

(5.14)

where $C_x = (r_b/B)/B$.

The plot of (5.14) is shown in Figure 5.9 for variable number of users, l. We observe in Figure 5.9 that the larger the number of users, the smaller the allocated channel capacity with increased noise tolerance level. However, for a large number of users we can use well-known power series approximation, that is, $\log_e(1+x) \le x$, to estimate the total capacity, where $-1 \le x \le 1$. In view

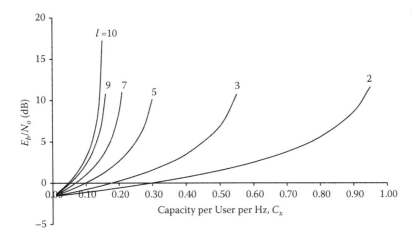

FIGURE 5.9
Normalized capacity of noncooperative CDMA system.

of (5.13) and upon application of the well-known power series approximation, we can write

$$C_x \leq C_x \left\{ \frac{E_b/N_o}{1 + C_x(l-1)E_b/N_o} \right\} \log_e \quad (5.15)$$

Alternatively,

$$C_x \leq \log_2 e - \frac{1}{E_b/N_o}$$

$$\leq \frac{1}{\log_e 2} - \frac{1}{E_b/N_o}$$

$$< \frac{1}{\log_e 2} \quad (5.16)$$

The preceding expressions and Figure 5.9 show that the total capacity does
not increase with number of users as in FDMA and TDMA systems.

5.2.2 Case II: Cooperative CDMA

In the cooperative CDMA case, the receiver for each user code or signal does
know the spreading waveforms of the other users. As a consequence, the
system receiver comprising a bank of l single-user receivers knows all the

l users' spreading waveforms and jointly detects and demodulates all users' codes. If we assign each user:

A transmission rate R_{ri} (where $i = 1, 2, \ldots l$)

A codebook containing a set of ($2^{k(L_c-1)}$) codewords[*] of carrier power C_l
(where k = number of input bits and L_c = constraint length[†])

then the achievable *l*-dimensional rate region for the *l* users may be represented by

$$R_r = \sum_{i=1}^{l} R_{ri} < B \log_2 \left(1 + \frac{lC_i}{N} \right) \qquad (5.17)$$

Note that this expression assumes equal carrier power $C_i = C$ for each user. The transmission bit rate is

$$R_r = r_b \frac{T_s}{T_b} \text{ bit/s} \qquad (5.18)$$

where r_b = input digital bit rate of the user (bit/s), T_s = symbol (or frame) duration (s), and T_b = bit (or burst) duration (s).

It follows from this expression that if the rates R_{ri} were selected to fall within the specified capacity region, the probabilities of errors[‡] for *l* users tend to zero as the code block-length tends to infinity. Suffice it to say that the sum of the rates tends to infinity with *l* users. In this instance, the cooperative CDMA has a similar format as the FDMA and TDMA.

Further comparison with FDMA and TDMA systems shows that if all the rates in the CDMA are selected to be identical, then (5.17) reduces to

$$R_r < B \log_2 \left(1 + \frac{lC_i}{N} \right) \qquad (5.19)$$

which incidentally is identical to the constraint applied to TDMA and FDMA, noting that $N = BN_o$ and lC_i is the total carrier power for all the users. We conclude that if the aggregated rates are unequally selected, there exists a point where the rates in the CDMA system exceed the capacity of FDMA and TDMA. This point will be fun for the reader to establish.

---

[*] A codeword is the number of bits transmitted in a block. For example, if we transmit *n* bits in a block, which is made up of *k* message bits plus *r* parity check bits, then the *n*-bit block is called a codeword.

[†] Constraint length is defined as the number of shifts over which a single message bit can influence the encoder output. This definition becomes clearer in Chapter 6.

[‡] As defined in Chapter 3, Equation (3.12).

5.3 Multiple-Access Protocols

The problem of multiple access (i.e., how to coordinate the access of multiple sending and receiving nodes to a shared broadcast channel, which essentially is that of *conflict resolution*) involves the sharing of a single communication channel among a group of geographically distributed stations, nodes, or users. The distributed algorithm that determines how stations, nodes, or users share a channel—that is, determining when a station can transmit—is known as the *multiple-access protocol*. In essence, the multiple-access protocol is the enabling satellite resource sharing or task scheduling procedure that prevents two or more nodes from transmitting at the same time over a broadcast channel. If they do, we have a collision, and receivers will not be able to interpret the signal.

Channels can be either (1) fixed assigned or (2) demand assigned to stations, nodes, or users.

A typical solution to resolving the *fixed-assignment multiple-access* (FAMA) problem is to assign time for each user for broadcast by TDMA. In FAMA, the channel assignment is tightly controlled, irrespective of traffic changes, which, in terms of capacity, can be wasteful when the traffic is asymmetric. A FAMA protocol can also be implemented by the FDMA technique. When using FDMA, no coordination or synchronization is required among stations. Each station uses its own bandwidth without interference. A drawback is that other stations cannot use the allotted bandwidth when the allotted station is idle; it is inflexible, and lacks flexibility and reconfigurability. It should be noted that neither FDMA nor TDMA allows any time overlap of the user/station transmissions. It thus leaves us to consider CDMA or SSMA—which can allow overlaps of transmissions in both the frequency and time—or hybrid-access types as possible candidates for conflict-free protocol. This CDMA conflict-free property can be achieved by using quasi-orthogonal signals in conjunction with matching filters at the receiving stations. Its main advantage is the *capture* in asynchronous SSMA [11]. The FAMA protocols are the most effective techniques for satellite networks composed of a small number of stations, typically less than 10, with stable and predictable traffic patterns. In situations where the traffic pattern is random and unpredictable, fixed allocation of the channel bandwidth leads to inefficient use of transponder capacity. It is thus desirable to design multiple-access protocols that allocate capacity on demand in response to the station request for capacity.

The *demand assignment* problem can be solved with either a central controller or a distributed control algorithm [12]. The demand assignment technique (called *demand assigned multiple access* [DAMA]) attempts to match users' demands to available satellite capacity. For instance, satellite channels may be grouped together as an entity (or bulk asset), which allows DAMA to assign users/stations variable time slots matching their information transmission requirements. DAMA can be divided into *reservation* and *token passing* protocols. Dynamic allocation using reservation based on demand increases the transmission throughput. The reservation process can be *implicit* or *explicit* [13].

In the explicit reservation protocol, a portion of the channel capacity is used as a reservation subchannel, usually in the form of a reservation subframe, in which the users send reservation packets. As the multiple-access protocol for the reservation subchannel, most of the reservation protocols adopt either *contention* (e.g., slotted-Aloha, S-Aloha) or a fixed assigned TDMA protocol [16]. In networks with a large number of stations, contention is used to keep the number of reservation slots small. The boundary between the control subframe and the data subframe can be movable, which expands the control subframe to fill the unused frame time, reducing the actual contention for the control slots [14].

In the implicit reservation protocol, we neither establish the reservation subchannel nor use reservation packets. By implicit reservation protocol, we usually mean reservation-Aloha (R-Aloha), where a slot that was successfully used by a user/station in the preceding frame is reserved for the same user/station in the succeeding frame, that is, implicit reservation [15]. The reservation protocols in general are often employed in satellite networks.

In token passing protocol, each node or station receives a guaranteed share of the network bandwidth [16]. For example, when a node or station receives a token—a control packet that provides the right of access—it has the control of the channel, indicating that the node can transmit for a specific time. Once the station finishes transmitting or its access time has expired, the token is passed to the next station on the logical ring. The protocol consists of alternating data transmission and token passing phases. DAMA is most effective where there are multiple users operating at low to moderate duty cycles, which is the typical military usage pattern. DAMA protocols, when integrated with random access satellite channels' sensing technique for both TDMA and CDMA techniques, have the potential to cope with more predictive and heavy traffic [17].

5.4 Summary

This chapter discussed the techniques that permit several earth terminals or stations on the same network to exchange information via a nodal point that the satellite represents. These techniques are FDMA, TDMA, and CDMA; they exploit the geometric advantage created by the use of the satellite. Comparisons between the capacities of these multiple-access methods were examined in terms of the information rate that each method achieves in an ideal additive white Gaussian noise channel of finite bandwidth. The principles of multiple-access protocols were explained.

Problems

1. What does the phrase "access to a communication satellite" mean?
2. Explain the principle behind the following schemes: FDMA, CDMA, and TDMA. Is it operationally possible to have a combination of these schemes? Give examples.
3. A TDMA slot acquisition system uses a frequency burst of 20 μs to determine its range. The satellite has a 500-MHz bandwidth and can produce a 10-dB downlink carrier. How much lower in power can the frequency burst be, relative to a TDMA carrier, to produce the same carrier-to-noise ratio in its filter bandwidth?
4. A QPSK TDMA system is to transmit 10,000 digital voice channels. Each channel has 4 bits per sample at a rate of 64 kb/s. It is envisaged that the system will accommodate 100 data bits/slot, at a frame efficiency of 15%. Determine:
   a. The satellite bandwidth required.
   b. The number of slots per frame.
   c. The number of preamble bits to be used.

5. In the FDMA system, frequency band selection may be fixed or assigned when demanded. Describe the difference between these two schemes. What action must be taken to prevent carrier cross talk?

References

1. Spilker, J.J. (1977). *Digital communications by satellite.* Englewood Cliffs, NJ: Prentice-Hall.
2. Puente, J.G., and Werth, A.M. (1974). Demand-assigned service for the Intelsat Global Network. *IEEE Spectrum*, 8(1), 59–69.
3. Gargliardi, R.M. (1991). *Satellite communications.* New York: Van Nostrand Reinhold.
4. Tri, T.H. (1986). *Digital satellite communications.* New York: McGraw-Hill.
5. Frenkel, G. (1974). The grade of service in multiple-access satellite communications systems with demand assignments. *IEEE Transactions on Communications*, COM-22, 1681–1685.
6. Maral, G., and Bousquet, M. (1993). *Satellite communications systems.* New York: John Wiley.
7. ITU-R (1988, 2012).

8. Viterbi, A.J. (1995). *CDMA: principles of spread spectrum communication.* Reading, MA: Addison-Wesley.
9. Shannon, C.E. (1948). A mathematical theory of communication. *Bell System Technical Journal*, 27.
10. Bamisaye, A.J., and Kolawole, M.O. (2010). Capacity and quality optimization of CDMA networks. *Journal of Telecommunications and Information Technology*, 4, 101–104.
11. Lam, S.S. (1979). Satellite packet communication—multiple access protocols and performance. *IEEE Transactions on Communications*, COM-27, 1456–1466.
12. Pritchard, W.L. (1977). Satellite communication—an overview of the problems and programs. *Proceedings of IEEE*, 65.
13. Retnadas, R. (1980). Satellite multiple access protocols. *IEEE Communications Magazine*, 18(5), 16–22.
14. Peyravi, H. (1999). Medium access control protocols performance in satellite communications. *IEEE Communications Magazine*, 62–71.
15. Tasaka, S. (1986). Performance analysis of multiple access protocol. In *Computer systems series, research reports and notes* (Schwetman, H., ed.). Cambridge, MA: MIT Press.
16. Stallings, W. (1991). *Data and computer communications.* New York: Macmillan.
17. Gaudenzi, R.D., and Herrero, O.R. (2009). Advances in random access protocols for satellite networks. *Proceedings of IEEE International Workshop on Satellite and Space Communications*, 331–336.

6

Error Detection and Correction for Coding Schemes

Coding techniques are used for several reasons: reduction of dc (direct current) wandering, suppression of intersymbol interference, and self-clocking capability. These problems are of engineering interest in the transmission of digital data. *Error correction* and *error detection* coding—collectively called *error control coding*—is the means whereby errors introduced into transmitted data can be detected and corrected at the receiver.

The coding schemes discussed in this chapter are for error detection and correction, specifically *forward error correction* (FEC) techniques. FEC codes may be divided into two classes of codes: linear block codes and convolutional codes [1]. Depending on the intended use of codes, FEC enables the receiver not only to detect errors but also to facilitate their correction.

To enhance the reader's understanding of the subject matter, we briefly introduce the principle of channel coding, followed by coding and decoding schemes. This arrangement enables the reader to move on to concrete realizations without great difficulty.

6.1 Channel Coding

Noise within a transmission channel inevitably causes discrepancies or errors between the channel input and channel output. A channel is a specific portion of information carrying capacity of a network interface specified by a certain transmission rate. In Chapter 3, Section 3.2.5 and Chapter 5, Section 5.2, the concepts of bit error rate and channel capacity using Shannon's [2] theorem were respectively discussed in some detail.

Coding enables the receiver to decode the incoming coded signals not only to detect errors but also to correct them. Effectively encoding signals from discrete sources and approximating a specified Shannon limit is achievable by symbol combination in large blocks and by the application of special coding methods such as Huffman [3] codes and Fano* [4] codes. For clarification, the terms *data compression* and *effective encoding* are treated as synonyms,

---

* Also called the Shannon-Fado codes.

meaning that redundancy is eliminated from the original signal, and it is then encoded as economically as possible with a minimum number of binary symbols. An encoder performs redundancy elimination. The application of the effective encoding methods essentially reduces the required channel transmission rate [5].

In general, if k binary digits enter the channel encoder, and the channel encoder outputs n binary digits, then the code rate can be defined as

$$R_c = \frac{k}{n} \tag{6.1}$$

The channel coding theorem specifies the capacity C_c as a fundamental limit on the rate at which the transmission of a reliable error-free message can take place over a *discrete memoryless channel* (DMC). The concept of DMC has been discussed in Chapter 3. For example, let a discrete memoryless source with an alphabet X have entropy $H(X)$ and produce one symbol every T_s seconds. Supposing that a DMC with capacity C_c (bit/symbol) is used once every T_c seconds, then if

$$\frac{H(X)}{T_s} \leq \frac{C_c}{T_c} \tag{6.2}$$

there exists a coding scheme for which the source output can be transmitted over the channel and be retrieved and reconstructed with an arbitrarily small probability of error. Conversely, it is impossible to transfer information over the channel and reconstruct the source signal with an arbitrarily small probability of error if

$$\frac{H(X)}{T_s} > \frac{C_c}{T_c} \tag{6.3}$$

The parameter C_c/T_c is called the *critical rate*. When $H(X)/T_s = C_c/T_c$, the system is said to be operating at the critical rate (at capacity). Figure 5.8 (in Chapter 5) has demonstrated that operating a system at capacity is achievable only if the value of E_b/N_o is above the Shannon limit; that is, $\log_e 2$ (or -1.6 dB). Recently, however, parallel-concatenated convolutional codes have achieved a performance close to this theoretical limit [6] (more is said of the convolutional coding technique in Section 6.2.2). In general, the objective is to achieve maximum data transfer in a minimum bandwidth while maintaining an acceptable quality of transmission. To achieve this, an error-detecting code must be devised. The most desirable solution is to use a forward error correction (FEC) technique. Error detection coding is a technique for adding redundant (extra) bits to a data stream in such a way that an error in the data stream can be detected [7].

6.2 Forward Error Correction Coding Techniques

Forward error correction (FEC) techniques are widely used in control and telecommunication systems. FEC codes may be divided into two broad types: linear block codes and convolutional codes.

6.2.1 Linear Block Codes

A block code is any code defined with a finite codeword length n, as seen in Figure 6.1.

Suppose the output of an information source is a sequence of binary digits 0 or 1. In linear block codes, the encoder splits up (or maps) the incoming data stream into blocks of k digits and processes each block individually by adding redundancy (extra digits) as parity check according to a predefined algorithm. The output of the encoder is a *codeword* with n digits, where $(n > k)$. The code rate is also given by (6.1).

A desirable property for a linear block code to possess is the *systematic structure* of the codewords as shown in Figure 6.1(b), where a codeword is divided into two parts: the message part and the parity part. The message part consists of k unaltered information (or message) digits, and the redundant checking part consists of $n - k$ *parity-check* digits, which are *linear sums* of the information digits. A codeword constructed in this way by appending parity symbols or bits to the payload data field is called a *systematic* code (or *linear systematic block code*).

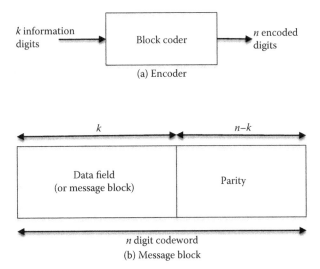

(a) Encoder

(b) Message block

FIGURE 6.1
A block code: (a) encoder; (b) message block.

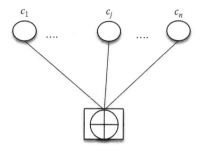

FIGURE 6.2
A representation of a parity equation.

For clarity, a parity equation can be explained using Figure 6.2. This is an equation that links n binary data c_i to each other by the *exclusive or, XOR,* logic, denoted by the $\oplus$ operator, where $i = 1, 2, ..., n$. From Figure 6.2, the parity equation is

$$c_1 \oplus c_2 \oplus c_3 \oplus \cdots c_{n-1} \oplus c_n = 0 \qquad (6.4)$$

This expression is called *parity constraint*, and is satisfied if the total number of 1s is even or null.

Example 6.1

Consider a string of encoded data 1010011 with a parity-check code sent through a noisy channel. A data string $\mathbf{y}$ = [10010010] was received. Check whether the received data are a valid codeword.

SOLUTION

Using the parity-check [Equation (6.4)], we check the validity of the codeword; thus,

$$y_1 \oplus y_2 \oplus y_3 \oplus y_4 \oplus y_5 \oplus y_6 \oplus y_7 \oplus y_8 = 1 \oplus 0 \oplus 0 \oplus 1 \oplus 0 \oplus 0 \oplus 1 \oplus 0 = 1$$

The received codeword $\mathbf{y}$ is not a valid codeword, given that the sum is 1, indicating that the parity-check equation is not satisfied. This shows that at least one error occurred during the transmission. Linear block codes are built from this simplest elementary code: the single parity-check code. More is said about linear codes.

From the preceding discussion, and without loss of generality, we could say that a binary block code is *linear* if and only if the mod-2 sum of two codewords is also a codeword. This is a vital property that allows encoder designers to find a *generator matrix* that describes the code.

The general form of a linear block codeword is

$$\mathbf{Y} = G\mathbf{X} \qquad (6.5a)$$

where G is a generator matrix that creates the check bits from the data bits, while the input message vector $\mathbf{X}$ and code (output) vector $\mathbf{Y}$ have components defined as

$$\mathbf{X} = (x_1, x_2, x_3, \ldots, x_k)$$

$$\mathbf{Y} = (y_1, y_2, y_3, \ldots, y_n)$$

(6.5b)

From elementary mathematics, multiplication of matrices can be performed only when the number of columns in Y equals the number of rows in the matrix X. Hence, the generator matrix G will have k rows by n columns:

$$G = \begin{bmatrix} g_{11} & g_{12} & \cdots & g_{1n} \\ g_{21} & g_{22} & \cdots & g_{2n} \\ \vdots & \vdots & \cdots & \vdots \\ g_{k1} & g_{k2} & \cdots & g_{kn} \end{bmatrix}$$

(6.6)

The addition that occurs in the matrix multiplication is mod-2 arithmetic; this becomes clearer in the example that follows.

Example 6.2

Consider message data X to be a 2-bit data word and the generator matrix [8]

$$G = \begin{bmatrix} 1 & 0 & 1 \\ 0 & 1 & 1 \end{bmatrix}$$

(6.7a)

Then the codeword Y is

$$Y = \begin{bmatrix} 1 & 0 & 1 \\ 0 & 1 & 1 \end{bmatrix} \begin{bmatrix} 1 & 0 \end{bmatrix}$$

$$= [(1 \otimes 1) \oplus (0 \otimes 0)][(1 \otimes 0) \oplus (0 \otimes 1)][(1 \otimes 1) \oplus (0 \otimes 1)]$$

$$= (1 \oplus 0)(0 \oplus 0)(1 \oplus 0)$$

$$= 101$$

(6.7b)

Note that operators $\otimes$ and $\oplus$ denote *logical conjunction*, AND logic and *exclusive or*, XOR logic, respectively. Thus, Y is a (3, 2) codeword having 2 data bits and 1 parity bit.

The process of code design is the process of determining the elements of the generator matrix. The subject of code design is involved, and the interested reader should consult Lin and Costello [7] for detailed treatment of the subject, and Shamnugam [9] for a review of coding for communication purposes.

Without loss of generality, the code in Example 6.2 can be considered systematic. A code is systematic if the first k bits of codeword Y constitute the message sequence X. Thus, a systematic code, in general, will have a generator matrix of the form:

$$G = \begin{bmatrix} 1 & 0 & \cdots & 0 & g_{1,k+1} & g_{1,k+2} & \cdots & g_{1,n} \\ 0 & 1 & \cdots & 0 & g_{2,k+1} & g_{2,k+2} & \cdots & g_{2,n} \\ \vdots & \vdots & \cdots & \vdots & \vdots & \vdots & \cdots & \vdots \\ 0 & 0 & \cdots & 1 & g_{k,k+1} & g_{k,k+2} & \cdots & g_{k,n} \end{bmatrix} \tag{6.8}$$

or

$$G = \begin{bmatrix} I_k & \vdots & P \end{bmatrix}_{k \times n} \tag{6.9}$$

where I_k is the k by k identity matrix. P is an arbitrary k by $(n - k)$ dimensional matrix, which represents the last $(n - k)$ of the generator matrix.

We are now ready to look at a technique of detecting errors in the codewords, which is accomplished by the use of a *parity-check* matrix H_Δ. In fact, the parity check is closely related to the generator matrix by

$$H_\Delta = \begin{bmatrix} g_{1,k+1} & g_{1,k+2} & \cdots & g_{1,n} \\ g_{2,k+1} & g_{2,k+2} & \cdots & g_{2,n} \\ \vdots & \vdots & \cdots & \vdots \\ g_{k,k+1} & g_{k,k+2} & \cdots & g_{k,n} \\ 1 & 0 & \cdots & 0 \\ \vdots & \vdots & \cdots & \vdots \\ 0 & 0 & \cdots & 1 \end{bmatrix}$$

$$= \begin{bmatrix} P \\ \cdots \\ I_{n-k} \end{bmatrix} \tag{6.10}$$

It is worth noting that for systematic codes the parity check can be written by inspection from the generator matrix.

Error detection is achieved by multiplying a received codeword Z, which has been corrupted, by the parity-check matrix H_Δ. This is a process of decoding a linear block code. The multiplication of Z and H_Δ is called *error syndrome*, S:

$$S = Z H_\Delta \tag{6.11}$$

The syndrome is a single word of length $(n - k)$: the number of parity-check bits in the codeword. If the codeword Z is correct, $Z = Y$, implying that

$$Y H_\Delta = [0] \tag{6.12a}$$

and

$$S = ZH_\Delta = [0] \qquad (6.12\text{b})$$

If the codeword Z is in error, that is,

$$Z = Y + \varepsilon \qquad (6.13)$$

the error in the transmission codeword will be detected if ε is not zero, where ε is an *error vector*. So,

$$Z = YH_\Delta$$
$$= [Y \oplus \varepsilon] H_\Delta \qquad (6.14)$$
$$= YH_\Delta \oplus \varepsilon H_\Delta$$

Since YH_Δ from (6.12a) is zero, it follows that error syndrome

$$S = \varepsilon H_\Delta \qquad (6.15)$$

If the syndrome S is nonzero, we know that an error has occurred in transmission of the codeword. Conversely, if the syndrome S is zero, Z is a valid codeword. Of course, S may also be zero for some combinations of multiple errors. Efficient decoders use syndrome to represent the error pattern, which can then be corrected. A more elaborate systematic parity-check code is the *Hamming code*. The Hamming code is an example of error detection properties of linear block codes.

Example 6.3

Consider a (7, 4) Hamming code. This code has 4 message bits and 3 parity-check bits in a 7-bit codeword. The generator matrix G for this code is given by

$$G = \begin{bmatrix} 1 & 0 & 0 & 0 & 1 & 1 & 1 \\ 0 & 1 & 0 & 0 & 1 & 1 & 0 \\ 0 & 0 & 1 & 0 & 1 & 0 & 1 \\ 0 & 0 & 0 & 1 & 0 & 1 & 1 \end{bmatrix} = \begin{bmatrix} I_4 & \vdots & P_3 \end{bmatrix} \qquad (6.16)$$

Using (6.10), the parity-check matrix can be formed:

$$H_\Delta = \begin{bmatrix} 1 & 1 & 1 & 0 & 1 & 0 & 0 \\ 1 & 1 & 0 & 1 & 0 & 1 & 0 \\ 1 & 0 & 1 & 1 & 0 & 0 & 1 \end{bmatrix} = \begin{bmatrix} P_4 & \vdots & I_3 \end{bmatrix} \qquad (6.17)$$

If we consider a message $X = (1010)$, the corresponding codeword Y, generated from (6.5a),

$$Y = GX = \begin{bmatrix} 1 & 0 & 0 & 0 & 1 & 1 & 1 \\ 0 & 1 & 0 & 0 & 1 & 1 & 0 \\ 0 & 0 & 1 & 0 & 1 & 0 & 1 \\ 0 & 0 & 0 & 1 & 0 & 1 & 1 \end{bmatrix} [1010] = 1010010 \qquad (6.18a)$$

The syndrome

$$S = YH_\Delta = [1010010] \begin{bmatrix} 1 & 1 & 1 & 0 & 1 & 0 & 0 \\ 1 & 1 & 0 & 1 & 0 & 1 & 0 \\ 1 & 0 & 1 & 1 & 0 & 0 & 1 \end{bmatrix} = 000 \qquad (6.18b)$$

If instead of the valid received codeword Y, an error is inserted into the second bit by changing the 0 bit to 1 bit, and the corrupted codeword is written as

$$Z = 1\underline{1}10010 \qquad (6.19a)$$

where the underlined bit is the error bit, then the syndrome of the newly received codeword is

$$S = \begin{bmatrix} 1 & 1 & 1 & 0 & 0 & 1 & 0 \end{bmatrix} \begin{bmatrix} 1 & 1 & 1 \\ 1 & 1 & 0 \\ 1 & 0 & 1 \\ 0 & 1 & 1 \\ 1 & 0 & 0 \\ 0 & 1 & 0 \\ 0 & 0 & 1 \end{bmatrix} = 110 \qquad (6.19b)$$

Since S is nonzero, two errors have been detected, indicating that the codeword is incorrect.

Example 6.4

If the message word is the same as in Example 6.3 for the (7, 4) Hamming code, but received corrupted codewords (1) 1011011 and (2) 1110100, determine whether these errors can be detected.

SOLUTION

Comparing (1) and (2) with the message X in Example 6.3 shows that received message (1) has one error, while (2) has three errors inserted. So their syndromes are as follows:

$$S = \begin{bmatrix} 1 & 0 & 1 & 1 & 0 & 1 & 1 \end{bmatrix} \begin{bmatrix} 1 & 1 & 1 \\ 1 & 1 & 0 \\ 1 & 0 & 1 \\ 0 & 1 & 1 \\ 1 & 0 & 0 \\ 0 & 1 & 0 \\ 0 & 0 & 1 \end{bmatrix} = 010 \quad (6.20)$$

An error has been detected since $S \neq [0]$, indicating that the codeword is invalid.

$$S = \begin{bmatrix} 1 & 1 & 1 & 0 & 1 & 0 & 0 \end{bmatrix} \begin{bmatrix} 1 & 1 & 1 \\ 1 & 1 & 0 \\ 1 & 0 & 1 \\ 0 & 1 & 1 \\ 1 & 0 & 0 \\ 0 & 1 & 0 \\ 0 & 0 & 1 \end{bmatrix} = 000 \quad (6.21)$$

The syndrome is zero. The errors are not detected because this is a legitimate codeword in the (7, 4) set. This demonstrates that the (7, 4) Hamming code can always reliably detect up to two errors; it also leads to the formulation of useful rules that tell us how many errors a given code can detect or correct. These rules define the capabilities of linear block codes in terms of *weight*, w_H; *distance*, d_H; and *minimum distance*, $d_{H,min}$.

Rule 1: The minimum distance (or Hamming distance) $d_{H,min}$ of a linear block code is the minimum weight (or Hamming weight), w, of any nonzero codewords, concisely written as

$$d_{H,min} = \min_{\substack{y \in Y \\ Y \neq 0}} w_H(Y) \quad (6.22)$$

The weight (or Hamming weight) of a code vector **Y** is the number of nonzero components of **Y**. The distance (or Hamming distance) between two code vectors $\mathbf{Y}_1$ and $\mathbf{Y}_2$ is the number of components by which they differ. Table 6.1 is used as an example of how the weights and distances are calculated. It can be seen in Table 6.1 that the minimum weight is 3 and the minimum distance is also 3.

Rule 2: The number of errors that can be detected in a code with minimum distance is

$$n_d = d_{H,min} - 1 \quad (6.23)$$

TABLE 6.1

Codewords and Weights for the (7, 4)
Hamming Code

Message Bits, X	Codeword, Y	Weight, w_H
0000	0000000	—
0001	0001011	3
0010	0010101	3
0011	0011110	4
0100	0100110	3
0101	0101101	4
0110	0110011	4
0111	0111000	3
1000	1000111	4
1001	1001100	3
1010	1010010	3
1011	1011001	4
1100	1100001	3
1101	1101010	4
1110	1110100	4
1111	1111111	7

The number of errors that can be counted to the *next-lowest* integer is

$$n_c = \frac{(d_{H,\min} - 1)}{2} = \frac{n_d}{2} \qquad (6.24)$$

For example, from Table 6.1:
The minimum distance $d_{H,min} = 3$.
The number of errors that can be detected $n_d = 2$.
The number of errors that can be counted $n_c = 1$.

Example 6.4 has demonstrated that detecting more than two bits that are in error becomes more difficult. This limits the use of the general rules for the linear block coding technique.

6.2.1.1 Cyclic Codes

Cyclic codes are a subset of linear block codes. They are the most useful and popular because encoding and decoding can be implemented by using simple shift registers: error detection and correction are achieved with shift registers and some additional logic gates. With a cyclic code, an end-around shift of a codeword yields another codeword. For example, if

$$(y_1, y_2, y_3, \cdots, y_{n-1}, y_n)$$

is a codeword, then, by definition, a cyclic code is also a codeword of the form

$$(y_n, y_1, y_2, y_3, \cdots, y_{n-1})$$

It follows from the above cyclic code description that only $n - 1$ codewords can be generated by cyclic shifts of a single codeword.

Encoding is achieved by using either a $(n - k)$-stage shift register or a k-stage shift register. Syndrome calculation for decoding is achieved by a $(n - k)$-stage shift register or a k-stage shift register. An example of encoding and decoding of a cyclic code is shown in Figure 6.3.

Figure 6.3(a) is a (7, 4) encoder. It generates a 7-bit codeword using a three-shift register (r_0, r_1, r_2). The message bits form the first 4 bits of the 7-bit codeword. When the gate is closed and the output switch is set to position 1, the most significant bits of the message bits are transmitted first. As the 4 message bits are transmitted, the gate is opened, allowing the data bits to be shifted into the three-shift register. After the 4 bits have been transmitted and fed into the shift register, the output switch is set to position 2. The contents of the shift register are then transmitted, forming the 3 parity-check bits.

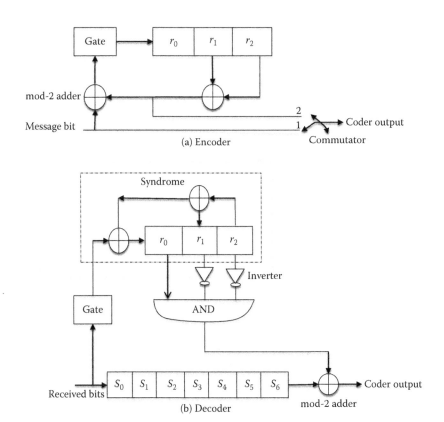

FIGURE 6.3
A (7, 4) cyclic code generator: (a) encoder; (b) decoder.

The output buffer from the encoder is the 7-bit-long codeword, which is fed into the syndrome forming part of the decoder in Figure 6.3(b). The syndrome output is AND gated, acting as the error detector. The AND gate gives a binary 1 output only when the AND input equals 100. If the AND output is 0, the current bit in the received codeword is correct and the received sequence is then shifted out of the decoder.

Cyclic codes have rich algebraic properties, which allow for efficient decoding techniques. A number of books, including Lin and Costello [7], Shamnugam [9], and Peterson [10], have provided excellent mathematical treatment of cyclic codes for which inquiring readers should consult.

6.2.1.2 BCH Codes

The Bose-Chaudhuri-Hocquenghem (BCH) codes [11] are perhaps the most powerful and flexible group of cyclic error correction codes available. Their design is straightforward: it uses shift register and logic circuits for coding and decoding. For instance, for a given block length n, codes can be designed with a number of gates and error correction capability. If e'_k is the number of correctable errors per codeword, the BCH codes would have the following properties:

Length of shift register: m'_k stages

Block length: $n = 2^{m'_k} - 1$, $m'_k > 2$ $\hspace{2cm}$ (6.25)

Minimum Hamming distance: $d_{H,\min} < 2e'_k + 1$ $\hspace{1.5cm}$ (6.26)

Number of parity-check bits: $n - k \le e'_k m'_k$ $\hspace{1.5cm}$ (6.27)

When error occurs in bursts, the number of parity-check bits needed to correct burst errors of length q has a lower bound given by

$$n - k \ge 2q \hspace{2cm} (6.28)$$

The code rate is

$$R_c = \frac{k}{n} = \frac{m'_k}{2^{m'_k} - 1} \hspace{2cm} (6.29)$$

Example 6.5

Consider a (7, 4) BCH code. In this instance, $n = 7$, $k = 4$. Using the above properties,

1. $e'_k = 1$
2. $m'_k = 3$ [from (6.25)]
3. $d_{H,min} = 3$ [from (6.26)]

These empirical results show that the Hamming code is a single-error correcting BCH code.

An important class of nonbinary BCH codes is the Reed-Solomon codes in which the symbols are blocks of bits [12]. Their importance is the existence of practical decoding techniques, as well as their ability to correct bursts of errors. NASA used Reed-Solomon codes for some of their missions to Mars.

A note on error correction techniques: Where a series of codewords have good short-burst error correction properties, such as terrestrial microwave links with short but deep fades, interleaving of bits or symbols provides an alternative technique for correcting the errors at the receiver. As an illustration, if we allow the individual bits of a given codeword to be spaced δ_c bits apart, a burst of length δ_c would corrupt only 1 bit in each of δ_c codewords. This single error can easily be rectified at the receiver, making the transmission impervious to longer error bursts.

Interleaving is the way a sequence of bits or symbols is arranged or ordered. For example, an *interleaver*, typically implemented as a *block interleaver*, is a rectangular matrix or an "orthogonal memory" that read in data symbols row by row (or line by line) and read out column by column (see Figure 6.4). By this process, bits or symbols that were adjacent on writing are spaced apart by the number of rows when reading.

The *de-interleaving* process is simply the inverse of the interleaving process, that is, writing in column by column and reading out row by row,

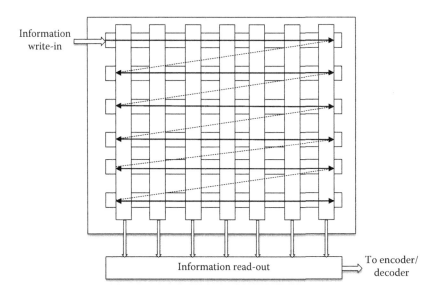

FIGURE 6.4
A block interleaver.

to achieve the original bit or symbol sequencing. In essence, by interleaving, correlated errors that might have been introduced in the transmission channel appear to be statistically independent at the receiver, thereby allowing better error correction. Block interleaving is simple to implement, and is used to overcome correlated burst error or fading in communication systems.

An interleaver could be a *row-column*, *helical*, *odd-even*, or *pseudorandom*, to name a few. Using Figure 6.5 as a guide, for an input x_i ($i = 1, 2, ..., 15$) we can describe [13]:

Input x_i				
x_1	x_2	x_3	x_4	x_5
x_6	x_7	x_8	x_9	x_{10}
x_{11}	x_{12}	x_{13}	x_{14}	x_{15}

Row-column interleaver output														
x_1	x_6	x_{11}	x_2	x_7	x_{12}	x_3	x_8	x_{13}	x_4	x_9	x_{14}	x_5	x_{10}	x_{15}

Helical interleaver output														
x_{11}	x_7	x_3	x_{14}	x_{10}	x_1	x_{12}	x_8	x_4	x_{15}	x_6	x_2	x_{13}	x_9	x_5

Odd-even interleaver output														
Encoder output without interleaving														
x_1	x_2	x_3	x_4	x_5	x_6	x_7	x_8	x_9	x_{10}	x_{11}	x_{12}	x_{13}	x_{14}	x_{15}
y_1	–	y_3	–	y_5	–	y_7	–	y_9	–	y_{11}	–	y_{13}	–	y_{15}
Encoder output with row-column interleaving														
x_1	x_6	x_{11}	x_2	x_7	x_{12}	x_3	x_8	x_{13}	x_4	x_9	x_{14}	x_5	x_{10}	x_{15}
–	z_6	–	z_2	–	z_{12}	–	z_8	–	z_4	–	z_{14}	–	z_{10}	–
Encoder final output														
y_1	z_6	y_3	z_2	y_5	z_{12}	y_7	z_8	y_9	z_4	y_{11}	z_{14}	y_{13}	z_{10}	y_{15}

FIGURE 6.5
Types of interleaver.

1. A *row-column* interleaver—the simplest—where data are written row-wise and read column-wise.
2. A *helical* interleaver is where the data are written row-wise and read diagonally.
3. An *odd-even* interleaver: In this type, the bits are left uninterleaved and encoded, but only the odd-positioned coded bits are stored. Next, the bits are scrambled and encoded, but now only the even-positioned coded bits are stored. Odd-even encoders can be used, when the second encoder produces 1 output bit per 1 input bit.
4. A pseudorandom interleaver is defined by a pseudorandom number generator or a lookup table.

6.2.1.3 Low-Density Parity-Check Codes

Since the introduction of a regular (3, 6) code with rate ½ of *low-density parity-check* (LDPC) codes by Gallager [14], where, within the node degree constraints, the parity-check matrix H_Δ is selected randomly, some LDPC code families have been built from circulants [15] to lower the descriptional complexity, including *accumulate-repeat-accumulate* (ARA) code [16], *progressive edge growth* (PEG) [17], and *multi-edge-type codes* [18]. The term *low density* comes from the fact that parity-check matrix H_Δ contains a large number of zeros (a lot more than half the elements of the matrix are zeros); as such, it is called a *sparse matrix*, e.g., Figure 6.6(a). A bipartite graph—also called *bigraph*

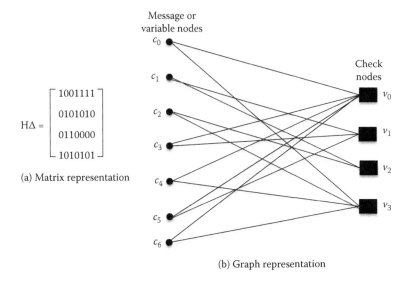

$$H\Delta = \begin{bmatrix} 1001111 \\ 0101010 \\ 0110000 \\ 1010101 \end{bmatrix}$$

(a) Matrix representation

(b) Graph representation

FIGURE 6.6
Representations of a *low-density parity-check* (LDPC) code: (a) matrix; (b) graph.

or Tanner graph [19]—has been used to represent the LDPC codes. A bipartite graph is a set of nodes decomposed into two disjoint sets such that no two nodes within the same set are adjacent. These nodes are called *check nodes* and *variable or message nodes*; an example is Figure 6.6(b), where each node is connected to a node in the other set. The construction of LDPC codes exists that approach Shannon's capacity to within hundredths of a decibel [20, 21].

An LDPC code parity-check matrix is called t-regular if the column weight (i.e., the number of 1s) for each column is exactly t, resulting in an average row weight of $nt/(n-k)$. If we fix the row weight to be exactly $s = nt/(n-k)$, then a (s, t)-regular LDPC has both the row and column weights fixed. As an illustration, with $t = 2$, the LDPC parity-check matrix is

$$H_\Delta = \begin{bmatrix} 1001111 \\ 0101010 \\ 0110000 \\ 1010101 \end{bmatrix} \tag{6.30}$$

and for any valid codeword c, we write

$$H_\Delta = \begin{bmatrix} 1001111 \\ 0101010 \\ 0110000 \\ 1010101 \end{bmatrix} \begin{bmatrix} c_0 \\ c_1 \\ c_2 \\ c_3 \\ c_4 \\ c_5 \\ c_6 \end{bmatrix} = 0 \tag{6.31}$$

Expression (6.31) serves as the starting point for constructing the decoder. By matrix/vector multiplication, Equation (6.31) defines a set of *parity checks*, which are written as

$$v_0 = c_0 \oplus c_3 \oplus c_4 \oplus c_5 \oplus c_6$$

$$v_1 = c_1 \oplus c_3 \oplus c_5$$

$$v_2 = c_1 \oplus c_2 \tag{6.32}$$

$$v_3 = c_0 \oplus c_2 \oplus c_4 \oplus c_6$$

Equation (6.32) is called the *parity-check constraints* equations. If the values assigned to the set of variable nodes represent a valid code, then each constraint equation is equal to zero, that is,

$$c_0 \oplus c_3 \oplus c_4 \oplus c_5 \oplus c_6 = 0$$

$$c_1 \oplus c_3 \oplus c_5 = 0$$

$$c_1 \oplus c_2 = 0 \tag{6.33}$$

$$c_0 \oplus c_2 \oplus c_4 \oplus c_6 = 0$$

From this example, that is, using Equations (6.30) and (6.32), we are in a position to explain encoding and decoding processes for LDPC codes.

6.2.1.3.1 *Encoding*

LPDC codes' encoding is performed in a similar way as in the linear codes earlier discussed. For instance, first reduce parity-check matrix H_Δ to systematic form H, that is, $H = [I_{n-k}|P]$, as demonstrated by (6.10), where I_{n-k} is an *identity matrix*. Once the systematic form of the parity-check matrix H is obtained, it is easy to confirm that a valid (systematic) generator matrix $G_s = [P^T|I_k]$ since $G_sH^T = 0$. As an example, consider the parity-check matrix H_Δ of (6.30), which can be reduced to systematic form as

$$H = \begin{bmatrix} 1000101 \\ 0100000 \\ 0010000 \\ 0001010 \end{bmatrix} \tag{6.34}$$

This expression determines the systematic generator matrix G_s.

It is worth noting that H, in general, may not have a fixed row or column weight, and, with high probability, P may be dense, which may make the encoder complex. While a good LDPC code is a good classical block code, a good classical code is not necessarily a good LDPC code. In the development of the LDPC codes, most critical is the sparsity of the parity-check matrix H_Δ, which is essential in keeping the decoding complexity low.

6.2.1.3.2 *Decoding*

The essence of a decoder is to reconstruct the transmitted codeword c from the possibly corrupted received word y by using the parity-check matrix H_Δ. For the decoded codeword to be the same as the transmitted (encoded) codeword, the parity-check constraints or equations must be satisfied, as expressed by (6.33).

LDPC code decoding is achieved through iterative processing based on the bipartite graph, to satisfy the parity-check constraints' conditions. LDPC decoders implement *message passing algorithms*, which are iterative, because at each round of the algorithms, messages are passed from message nodes to check nodes, and from check nodes back to message nodes. The messages from message nodes to check nodes are computed based on the *observed* value of the message node and some of the messages passed from the neighboring check nodes to that message node. For the observed data, the problem becomes that of inversion, which reverses the role of the data leading to the *likelihood function* or probability of occurrence given certain conditions. For instance, the message passed from a message node c to a check node v is the probability that c has a certain value given the observed value of that message node, and all the values communicated to c in the prior round from the check nodes incident to c other than v. Likewise, the message passed from check node v to variable node c is the probability that c has a certain value given all the messages passed to v in the previous round from message nodes other than c. In brief, this message passing process attempts to estimate the bit probabilities using a priori or *intrinsic* information (knowledge before an event) and a posteriori or *extrinsic* information (knowledge after an event). Through the repeated iterations of message passing along edges in the bipartite graph, decoding performance is achieved, with some stopping criterion.

6.2.1.3.3 *Likelihood Functions*

A detailed description of probability distribution functions, including the likelihood functions, is available in Kolawole [22]. However, to enable an understanding of the concept being developed in this section, and following [22], a brief description of likelihood functions follows.

Let the data vector $\mathbf{x} = \{x_1, x_2, x_3, ..., x_n\}$ be a random sample from an unknown population, and let $f(\mathbf{x}|w_i)$ specify the probability of observing data $\mathbf{x}$ given the parameter w_i of length k (i.e., $i = 1, 2, ..., k$). If individual observations are statistically independent of one another, we can write the probability density function for m individual observations as

$$f(x_1, x_2, x_3, ..., x_n|w_i) = f_1(x_1|w_i)f_2(x_2|w_i)f_3(x_3|w_i)...f_m(x_m|w_i)$$

$$= \prod_{j=1}^{m} f_j(x_j|w_i) \tag{6.35}$$

The likelihood function, $L(w_i|\mathbf{x})$ can be defined by reversing the roles of the data vector $\mathbf{x}$ and the parameter w_i in (6.35) as

$$L(w_i|\mathbf{x}) = f(\mathbf{x}|w_i) \tag{6.36}$$

A parameter that seeks the value of the parameter w_i that maximizes the likelihood function is called the *maximum likelihood estimation* (MLE)—an estimate that seeks the probability distribution that makes the observed data most likely. The MLE estimate is obtained by maximizing the log-likelihood function, that is, $\log L(w_i|\mathbf{x})$ (or $\ln L(w_i|\mathbf{x})$). The functions $L(w_i|\mathbf{x})$ and $\ln L(w_i|\mathbf{x})$ are monotonically related to each other; as such, the same MLE estimate can be obtained by maximizing either one. Suppose the log-likelihood function $\ln L(w_i|\mathbf{x})$ is differentiable; if $w_{i,MLE}$ exists, then it must satisfy at $w_i = w_{i,MLE}$:

$$\frac{\partial \ln L(w_i|\mathbf{x})}{\partial w_i} = 0 \tag{6.37}$$

which is known as the *likelihood equation*.

If the data are identically distributed, that is, possess identical statistical characteristics, then the data vector is said to be independently and identically distributed (i.i.d.). In this case the covariance matrix $cov[\mathbf{x}]$ becomes an identity matrix:

$$cov[\mathbf{x}] = E[\mathbf{x}\mathbf{x}^T] = \sigma^2 \boldsymbol{I} \tag{6.38}$$

where σ^2 is the variance of the components.

Like any self-organizing map or graph, Bayes' theorem provides the basis for acquiring knowledge about the *prior* probabilities of an alternative premise and about the probability of observing various data given the premise. Upon this acquisition, the Bayesian method then allows an a posterior probability to be assigned to each candidate premise based on these assumed *priors* and the observed data [22, 23].

Suppose that the probability $p(x_i)$ of values x_i is known, where $i = 1, 2, \ldots, n$. Suppose also that an event y occurs in conjunction with values x_i occurring. The question becomes how the event y has actually occurred, and what will its impact be on the individual probability of x_i? This translates to finding $p(x_i|y)$ for all values of i. By Bayes' theorem,

$$p(x_i|y) = \frac{p(x_i)p(y|x_i)}{p(y)} \tag{6.39}$$

Alternately,

$$p(x_i|y) = p(x_i)\frac{p(y|x_i)}{\sum_{i=1}^{n} p(y|x_i)p(x_i)} \tag{6.40}$$

where:

1. The first term, $p(x_i)$, is the initial or prior probability.
2. The second term, $\dfrac{p(y|x_i)}{\sum\limits_{i=1}^{n} p(y|x_i)p(x_i)}$, is the amended or posterior probability

   that corrects the *prior* probability on the basis of data in hand.
3. $p(y|x_i)$ = probability of observing y given that x_i holds.

Another metric that maximizes the likelihood is called *maximum a posteriori probability* (MAP). By Bayesian inference, the MAP estimate can be expressed thus:

$$\hat{x}_{i,MAP} = \underset{x_i}{\operatorname{argmax}}\, p(x_i|y) = \underset{x_i}{\operatorname{argmax}}\, p(x_i|y)p(x_i) \tag{6.41}$$

which implies the maximum x_i estimate with the highest posterior probability.
Also, the MAP estimate can be obtained by applying log trick:

$$\hat{x}_{i,MAP} = \underset{x_i}{\operatorname{argmax}}\left\{ \sum_{i=1}^{n} \log[p(x_i|y)] + \log[p(x_i)] \right\} \tag{6.42}$$

MAP is the direct counterpart to MLE.

Following the preceding a priori definitions, we can describe the codeword's status at and between the variable nodes and check nodes. As noted in the decoding introductory paragraph, we let y represent the possibly corrupted received where $y = (x_1, x_2, \dots x_n) + n_i$. In this instance, $(x_1, x_2, \dots x_n)$ denotes the codewords out of the LDPC encoder and n_i is additive noise, preferably Gaussian, with an i.i.d. where each component has variance σ^2. Given that the codewords are binary—0s and 1s—and $x_j = 0$ and $x_j = 1$ are transmitted, the decoding problem is to *estimate* valid codeword c's from the y's, by calculating the *log-likelihood ratio* (LLR) of the a posteriori or *extrinsic* information. That is,

$$LLR(x_j) = L(x_j) = \log\left(\frac{p(x_j = 0|y)}{p(x_j = 1|y)} \right) \tag{6.43}$$

for all j. Conditions are then set to evaluate the streamed codewords from the LLR estimates. For instance, if $LLR(xj) \geq 0$, the estimated codeword $\hat{x}_j = 0$; otherwise, the estimated codeword $\hat{x}_j = 1$.

Suppose, in Figure 6.6, there are M check nodes that correspond to the parity-check constraints and N variable nodes that represent the data bits of the codeword of the bipartite graph. The N code bits must satisfy all parity checks, thus enabling us to compute a posterior *probability* (APP), from the connected variables, which is $p(x_j = d|E_j, y)$, where $d = 0, 1$, and E_j is the event that all parity checks associated with x_j have been satisfied. Following (6.39), we write

$$p(x_j = d|E_j, y) = p(y|x_j = d)p(E_j|x_j = d, y) \qquad (6.44)$$

For the set of M check-parity nodes to code bit x_j, and for the set of N-bit variable nodes connected to the ith parity check, the standard *message passing algorithm* (MPA) decoding procedure is thus expressed.

1. Initialization: For each variable node, $j = 1, 2, \ldots, N$ with the received information y from the source, and calculate the initial log-likelihood ratio (LLR)—which is a posteriori—using (6.43):

$$L(x_j) = LLR(x_j) = \log \frac{p(x_j = 0|y_j)}{p(x_j = 1|y_j)} \qquad (6.45)$$

Suppose the received data are modulation over an additive white Gaussian noise (AWGN) channel with unity mean and variance σ^2; the LLR for every variable node can be represented as

$$\alpha_j^0 = L(x_j) = \frac{2y_i}{\sigma^2} \qquad (6.46a)$$

while messages along edges (to check nodes) are initialized to zero, that is,

$$\beta_j^0(x_j) = 0 \qquad (6.46b)$$

Check node LLR and messages (to variable nodes) are both initialized to zero.

For the nth iteration, we use the following notations to represent each node's participation in the message passing:

$\alpha_{i,j}^n$: The message sent from variable node j to check node i.

$\beta_{i,j}^n$: The message sent from check node i to variable node j.

$M(j) = \{i : H_{\Delta ij} = 1\}$: The set of parity checks in which variable node j participates.

$N(i) = \{j : H_{\Delta ij} = 1\}$: The set of variable nodes that participate in parity check i.

$M(j)\backslash i$: The set $M(j)$ with check node i excluded.

$N(i)\backslash j$: The set $N(i)$ with variable node j excluded.

2. Check nodes update: For each check node $i = 1, 2, \ldots, M$, calculate LLR and check-to-variable node messages based on the incoming messages from variable nodes, using

$$\beta_{i,j}^n = \operatorname{sgn}\beta_{i,j}^n * \left|\beta_{i,j}^n\right| \tag{6.47}$$

where

$$\operatorname{sgn}\beta_{i,j}^n = \prod_{j \in N(i)\backslash j} \operatorname{sgn}\left(\alpha_{ij}^{n-1}\right) \tag{6.48}$$

$$\left|\beta_{i,j}^n\right| = \bigotimes_{j \neq j}\left(\alpha_{i,j}^{n-1}\right) \tag{6.49}$$

$$\alpha_{i,j}^{n-1}(x_j) = \log\left(\tanh\left(\frac{|x_j|}{2}\right)\right) \quad xj \geq 0, n > 1 \tag{6.50}$$

The symbol $\otimes$ denotes logic AND, and sgn—called *signum*—is the sign of a real number defined as

$$\operatorname{sgn}(x) = \begin{cases} -1 & x < 0 \\ 0 & \text{for} \quad x = 0 \\ +1 & x > 0 \end{cases}$$

3. Variable nodes update: For each variable node $j = 1, 2, \ldots, N$, calculate LLR and variable-to-check node messages based on the incoming messages from check nodes, using

$$\alpha_{i,j}^n = \alpha_{i,j}^0 + \sum_{i \in M(j)\backslash i} \beta_{i,j}^n \tag{6.51}$$

4. Verify parity checks: For each bit $j = 1, 2, \ldots, N$, check if:

$\alpha_j^n \geq 0$, then estimate $\hat{x}_j = 0$

$\alpha_j^n < 0$, then estimate $\hat{x}_j = 1$

Also if $H_\Delta^T \hat{x} = 0$, the decoding process is finished with $\hat{x}$ as the decoder output; otherwise, go to step 2—check node processing.

Declare the decoder a failure if within a number of reasonable iterations, say, i_{nmax}, the algorithm does not end, and output an error message.

LDPC codes were recommended for inclusion in the Consultative Committee for Space Data Systems (CCSDS) telemetry channel coding standard for deep-space applications [24].

6.2.2 Convolutional Codes

A convolutional coder is a finite memory system. The name *convolutional* refers to the fact that the added redundant bits are generated by mod-2 convolutions. A generalized convolutional encoder is shown in Figure 6.7. It consists of an L-stage shift register, n mod-2 adders, a commutator, and a network of feedback connections between the shift register and the adders. The number of bits in the input data stream is k. The number of output bits for each k-bit sequence is n bits. Since n bits are produced at the output for each input of k bits, the code rate is still $R_c = k/n$, the same as Equation (6.1).

A very important parameter in the consideration of convolutional encoding is the *constraint* or *memory length*. The constraint length is defined as the number of shifts over which a single message bit can influence the encoder output. For example, if input message data are in groups of k bits and are fed into the L-stage shift register, then the register can hold (L/k) groups. Given that each group produces n output bits, the constraint or memory length is

$$L_c = (L/k)n \tag{6.52}$$

This expression also represents the memory time of the encoder. Examples 6.6 and 6.7 will give the reader a better feeling for the convolutional encoding operation.

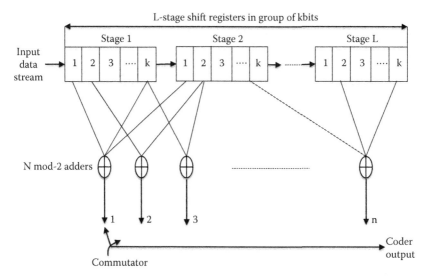

FIGURE 6.7
Convolutional codes encoder.

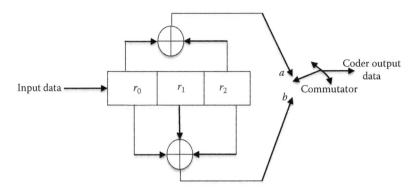

FIGURE 6.8
Rate 1/2 convolutional encoder.

Example 6.6

Consider a binary convolutional encoder, rate of 1/2, shown in Figure 6.8 with a constraint length of 3, output nodes $n = 2$, and one-digit input, $k = 1$. The shift register (r_0, r_1, r_2) is initially set to clear (i.e., all-zero state). The first bit of the input stream is entered into r_0. During this message bit interval the commutator samples, in turn, the adder outputs a and b. Consequently, a single bit yields, in the present case, two coded output bits, using the algebraic functions

$$a = r_0 \oplus r_2$$
$$b = r_0 \oplus r_1 \oplus r_2$$

(6.53)

Therefore, the encoder is of rate 1/2. The next message (input) bit then enters r_0, while the bit initially in r_0 transfers to r_1, and the commutator again samples the two mod-2 adder outputs. This process continues until eventually the last bit of the message (input) bits has been entered into r_0. Thereafter, to ensure that every input data bit may proceed entirely through the shift register, and hence be involved in the complete coding process, enough 0s are added to the message to transfer the last message bit through r_2, and thus out of the shift register. The shift register finds itself in its initial all-zero state (as in Table 6.2, column 1, shift 8). This may be verified by an example.

If the input stream to the encoder is a 5-bit sequence,

$$x = 10111$$

(6.54a)

then the coded output bit stream is (see Table 6.2)

$$ab = 11\ 11\ 10\ 00\ 01\ 10\ 01\ 00$$

(6.54b)

Alternative methods of describing a convolutional code are the tree diagram, the trellis diagram, and the state diagram. Example 6.7 will be used to explore these alternative methods.

TABLE 6.2

Table of Register Contents for Decoding

Shift	Input Bit	$r_0r_1r_2$	ab
1	1	100	11
2	0	010	11
3	1	101	10
4	1	110	00
5	1	111	01
6	0	011	10
7	0	001	01
8	0	000	00←reset

Example 6.7

Consider a binary convolutional encoder, rate of 1/3, shown in Figure 6.9. This figure is similar to Figure 6.8 except that the output v_1 is fed directly from r_0. For each message (input) bit, the sequence ($v_1\ v_2\ v_3$) is generated. It follows from Figure 6.9 that the output sequence is given by

$$v_1 = r_0$$

$$v_2 = r_0 \oplus r_2 \tag{6.55}$$

$$v_3 = r_0 \oplus r_1 \oplus r_2$$

Since the first bit in the output sequence is the message bit, this particular convolutional code is *systematic*. As a result, v_2 and v_3 can be viewed as parity-check bits.

The output sequence for an arbitrary input sequence is often determined by using a code tree. For example, the tree diagram for the above convolutional encoder is illustrated in Figure 6.10. Initially, the encoder is set to an

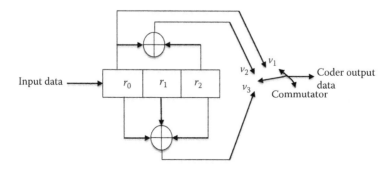

FIGURE 6.9
Rate 1/3 convolutional encoder.

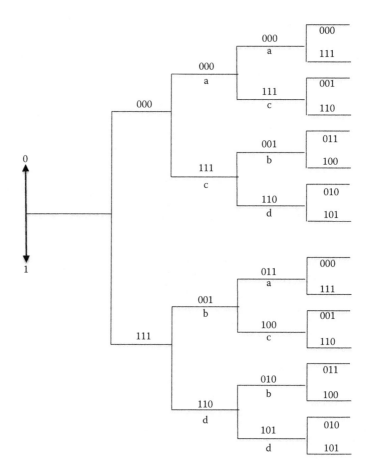

FIGURE 6.10
Code tree diagram of rate 1/3 coder of Figure 6.9. The branches correspond to input symbols. The branch upward corresponds to an input bit 0, whereas the downward branch corresponds to an input bit 1.

all-zero state. The tree diagram shows that if the first message (input) bit is 0, the output sequence is 000, and if the first input bit is 1, the output sequence is 111. If the first input bit is now 1 and the second bit is 0, the second set of 3 output bits will be 001. Continuing through the tree, we will be able to show that if the third input bit is 0, the output will be 011. If the third input bit is 1, the output will be 100. Supposing that a particular sequence takes us to a particular node in the tree, the branching rule allows us to follow the upper branch if the next input bit is 0 and the lower branch if the input bit is 1. Consequently, a particular path through the tree can be traced for a specific input sequence. It can be observed that the tree generated by this convolutional encoder (Figure 6.9) shows that the structure repeats itself after the third stage. The tree diagram is thus shown up to the third stage, as in Figure 6.10. This behavior is consistent with the fact that the constraint length is 3 ($L_c = 3$): interpreted as the 3-bit output sequence at each stage determined by an input bit and 2 bits contained in the first two stages (r_0, r_1) of the shift register.

It should be noted that the bit in the last stage (r_2) of the register is shifted out and does not affect the output, just as demonstrated in Table 6.2 for Example 6.6. In essence, it could be said that the 3-output bit for each input bit is determined by the input bit and four possible states labeled a, b, c, and d in Figure 6.10 and denoted, respectively, by 00, 01, 10, and 11. With this labeling, it can be observed in Figure 6.10 that, at the third stage:

There are two nodes each with the label a, b, c, or d.

All branches originating from two nodes and having the same label generate identical output sequences. This implies that two nodes having the same label can be merged.

By merging two nodes having the same label in the code tree diagram of Figure 6.10, another diagram emerges, as shown in Figure 6.11. This diagram is called the *trellis diagram*. The dotted lines denote the output generated by the lower branch of the code tree with input bit 1, while the solid lines denote the output generated by the upper branch of the code tree with input bit 0.

The completely repetitive structure of the trellis diagram in Figure 6.11 suggests that a further reduction is possible in the representation of the code to the *state diagram*. A state diagram, shown in Figure 6.12, is another way of representing states (a, b, c, and d) transitioning from one state to another.

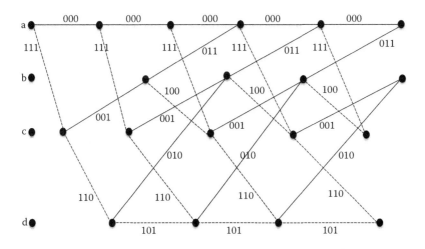

FIGURE 6.11
Trellis diagram of Figure 6.9. The solid line denotes the output generated by the input bit 0, whereas the dotted line denotes the output generated by the input bit 1. Labels a, b, c, and d denote the four possible states of the shift register.

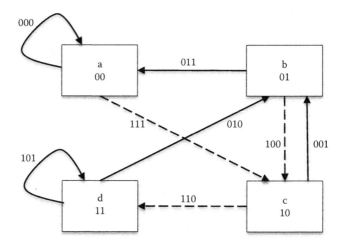

FIGURE 6.12
The state diagram of Figure 6.9. The solid lines denote that the input bit is 0, whereas the dotted lines denote that the input bit is 1. The three bits appended to each transitory line represent the output bits.

Arrows represent the transitions from state to state. The states of the state diagram are labeled according to the states of the trellis diagram. The 3 bits shown next to each transitory line represent the output bits.

From the preceding discussions, we are in a position to draw the code tree, trellis diagram, and state diagram for the fixed 1/2 rate convolutional coder of Figure 6.8. The message sequence 10111 is used as input to the encoder in Figure 6.8. The code tree of Figure 6.10 is drawn. The tree-drawing procedure is the same as described previously for the encoder in Figure 6.8 by moving up at the first branching level, down at the second and third, and up again at the fourth level to produce the outputs appended to the traversed branches. After the first three branches, the structure becomes repetitive, a behavior that is consistent with the constraint length of the encoder, which is 3 ($L_c = 3$). From the code tree, the trellis and state diagrams are drawn as shown in Figures 6.13 to 6.15, respectively.

Generalization of the preceding procedures can be made, without loss of credence, to code rate $R_c = k/n$. We have observed that the tree diagram will have 2^k branches originating from each branching node. Given that the effect of constraint or memory length L_c will be the same, paths traced for emerging nodes of the same label in the tree diagram will begin to remerge in groups of 2^k after the first L_c branches. This implies that all paths with $k(L_c - 1)$ identical data bits will merge together, producing a trellis of $2^{k(L_c-1)}$ states with all branchings and mergings appearing in groups of 2^k branches. Also, the state diagram will have $2^{k(L_c-1)}$ states, with each state having 2^k input branches coming into it. Thus, L_c can be said to represent the number of k-tuples stored in the shift register.

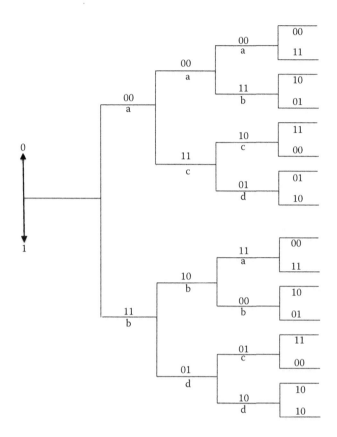

FIGURE 6.13

Code tree diagram of rate 1/2 coder of Figure 6.8. The branches correspond to input symbols. The branch upward corresponds to an input bit 0, whereas the downward branch corresponds to an input bit 1.

6.2.2.1 Decoding Convolutional Codes

As noted earlier, the function of a decoder is to keep track of the encoder's state transitions and reconstruct the input data stream. Conceptually, decoding is achieved by taking the received bits and finding the path through the code tree that lies closest in Hamming distance to the received bits. This approach might be impractical in the case of a long data sequence because a data stream of k symbols would require 2^k branches of the code tree to be searched for such a minimum Hamming distance. Among Huffman [3], Fano [4] (also called Shannon-Fano), and Viterbi [25, 26], the standard and most popular decoder is the *Viterbi algorithm*, discussed next. Many variations to, or generalizations of, the basic Viterbi algorithm have been reported in the literature. These variations include the weighted output Viterbi algorithm [27, 28], the soft-output Viterbi algorithm (SOVA) [29], the reliability estimation algorithm [30], and the list-type Viterbi algorithm (LVA) [31, 32].

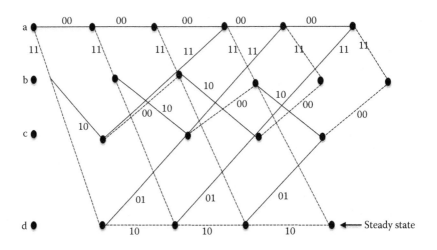

FIGURE 6.14
Trellis diagram of Figure 6.8. The solid line denotes the output generated by the input bit 0, whereas the dotted line denotes the output generated by the input bit 1. Labels a, b, c, and d denote the four possible states of the shift register.

All these variations attempt to transform path metric differences, which indicate the reliability of competing path sequences, into reliability values for individual symbols.

6.2.2.1.1 Viterbi Algorithm

In his paper, Viterbi [25] developed a convolutional decoding algorithm having a relatively short constraint length L_c. As noted earlier, a decoding

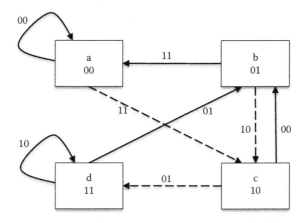

FIGURE 6.15
The state diagram of Figure 6.8. The solid lines denote that the input bit is 0, whereas the dotted lines denote that the input bit is 1. The three bits appended to each transitory line represent the output bits.

algorithm can be developed conceptually by taking the received sequence and finding the path through the code tree that lies closest in Hamming distance to the received sequence. After merging paths of equal metrics, only the surviving path and its metrics from the received sequence are stored. The Viterbi algorithm, to date, has been the most practical for convolutional coding.

A generalized Viterbi algorithm is described as follows. Denote a decoder's states as s, with its metric as $m(s, t)$ at any time t, where $s = a$, b, c, or d. Denote the Hamming distances for upper and lower branches entering a state as $d_U(s, t)$ and $d_L(s, t)$, respectively. A Hamming distance is defined by Equation (6.22). Initialize at $t = 0$ by setting all states' metrics to zero. Then:

1. For a given trellis node or state s and time t, compare the received codeword with each branch codeword entering the trellis node or state.

2. Calculate $d_U(s, t)$ and $d_L(s, t)$.

3. Modify the path by calculating D_U and D_L from the upper and lower branch distances using the previous and current state metrics:

$$D_u = d_U(s,t) + m(s_{t-1}, t-1)$$
$$D_l = d_L(s,t) + m(s_{t-1}, t-1)$$

(6.56)

where D_U and D_L are abstract terms.

4. Identify the surviving branch entering the state at time t considering:
   a. If $D_U < D_L$, select the upper branch as the survivor. Then write D_U as the final metric $m(s, t)$ for the state. Otherwise, select the lower branch as the survivor and write D_L as the final metric $m(s, t)$ for the state.
   b. If $D_U = D_L$, both paths are equal contenders; randomly select any branch as the survivor.
   c. Discard any branch that has not survived.

5. Repeat steps 1 to 4 until the end of the trellis tree at time $t = t_{rec}$, where t_{rec} is the time when the final trellis state has been reached.

6. Form all final state metrics, that is, $m(s, t_{rec})$.

7. Select the minimum distance (d_{min}) or metric, trace back the path from this state, and note the binary digits that correspond to branches, which is the path that has been traced back produced.

A word of caution! The Viterbi decoding algorithm does not work equally well with all possible convolution codes. Knowledge of the mechanisms involved in the Viterbi decoding failures can be used to design or select codes that perform well under Viterbi decoding [33].

In general, a Viterbi decoder is able to retain information on $2^{k(L_c-1)}$ paths for satisfactory performance [34]. For practical reasons, particularly to reduce the data-handling capacity required of the convolutional decoder, the trellis depth is approximated to $5L_c$ nodes. This allows decoding of only the oldest received sequence within the trellis and selecting the path with the fewest errors, while deleting the remaining paths.

6.2.2.2 Turbo Codes

A turbo code is a refinement of the concatenated encoding structure plus an iterative algorithm for decoding the associated code sequence. Berrou et al. [35] introduced turbo codes in 1993. Turbo codes are also called parallel concatenated convolutional code (PCCC) or serial concatenated convolutional code (SCCC). Since the introduction of this technique, many other classes of turbo code have been discovered, including serial versions (SCCC) and repeat-accumulate (RA) codes.

Turbo design encompasses two blocks: encoder and the decoder, as shown in Figure 6.16.

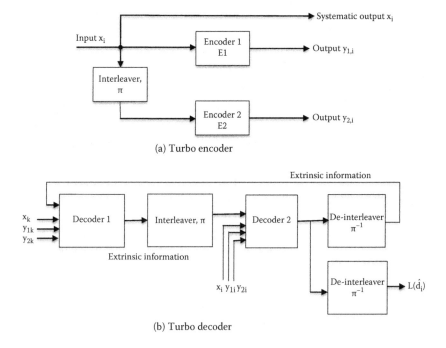

(a) Turbo encoder

(b) Turbo decoder

FIGURE 6.16
Basic turbo codes: (a) encoder; (b) decoder.

6.2.2.2.1 Turbo Encoding

In the encoder, as seen in Figure 6.16(a), the codes are constructed by using two or more component codes on different interleaved versions of the same information sequence. The encoder's output, consisting of message and parity bits, may be a multiplexed (and probably punctured) stream.

6.2.2.2.2 Turbo Decoding

In the decoder, as seen in Figure 6.16(b), there are two decoders for outputs from both encoders. Both decoders provide estimates of the same set of data bits, albeit in a different order. If all intermediate values in the decoding process are soft (i.e., real) values, the decoders can gain greatly from exchanging information, after appropriate reordering of values. Information exchange can be iterated a number of times to enhance performance (i.e., to produce more reliable decisions). At each round, decoders reevaluate their estimates, using information from the other decoder, and only in the final stage will hard decisions be made; that is, each bit $L(d_i)$ is assigned the value 1 or 0.

For convolutional codes, hard decoding means that an inner system has already made decisions about individual symbol values, specifically 1 or 0 is already decided. Soft decoding means that the inner system has made no decision but a decision is made on conditional probability values. Readers seeking a deeper understanding of making decisions based on likelihood functions should consult Chapter 6 of Kolawole [22].

Primarily, the decoders attempt to reestablish the transmitted data from the received systematic bit stream (including associated parity-check bit streams), and even though these may have been corrupted by noise, their decoding algorithm avoids passing hard decisions among them. It is worth noting that decoders need not initially make a hard decision about the transmitted symbol values. Instead, the decoders may calculate a measure of the relative likelihood or probability of the various input message values. This measure is called the *soft information*. A *soft-in-soft-out* (SISO) decoder receives as input a soft (i.e., real) value of the signal. The decoder then outputs for each data bit an estimate expressing the probability that the transmitted data bit was equal to 1.

Two main algorithms form the component of the turbo decoders:

1. The maximum a posteriori probability (MAP)—also called a SISO algorithm—maximizes the output probability based on some knowledge of the input data a priori *probabilities* and soft output (i.e., real value) from the demodulator [36]. Elements of MAP have been defined by Equations (6.41) and (6.42).

2. *Soft-output Viterbi algorithm* (SOVA). Hagenauer and Hoeher [29] introduced SOVA in 1989 as an augmentation to the usual Viterbi detector to provide additional estimates of the conditioned probability of a given symbol or bit value, at a given time, in the final surviving detected sequence. Basically, SOVA searches for the

maximum likelihood (ML) path in the code trellis, and records all "the metric difference between the two paths" values between the ML path and the paths remerging with it. The approximated logarithmic likelihood ratio (LLR) values can be updated by trace-back. LLR has been expressed by Equation (6.43).

Both of the algorithms use iterative techniques to achieve decoding performance.

Iterative turbo decoding methods have also been applied to more conventional FEC systems, including the Reed-Solomon/convolutional code structure.

6.3 Summary

The transmission of data over a communication link is highly likely to result in some errors occurring in the received data, for at least a small proportion of time, because of noise added by the transmission medium and system. Some techniques on how the errors can be detected and correction effected have been explored in this chapter. These techniques are forward error correcting, broadly classified under two headings: block codes and convolutional codes.

In the block coding techniques, the encoder splits up incoming data stream into blocks of finite-number digits and processes each block by adding extra bits (called redundancy) according to a predefined algorithm. The output of the encoder is a codeword with another finite number of digits. A few subsets of linear block codes, such as Hamming, cyclic, BCH, and LDPC, have also been discussed. Their importance is the existence of practical decoding techniques.

In the convolutional coding techniques, the encoder processes the incoming data stream continuously, while its decoder employs the Viterbi algorithm, among others. Convolutional codes are very popular because they are conceptually and practically simple to utilize.

Examples were sparingly used as illustrative tools to explain these forward error correction techniques.

Problems

1. A channel-coded CDMA system uses a rate 1/2, seven-stage convolutional encoder before addressing at each transmitter. Determine by which factor the total number of users in the system can be increased over the uncoded system. Also, determine the factor if the rate becomes 1/3.

2. Design a (7, 4) Hamming code that is different from Example 6.2.
   a. Provide the generator matrix of your design.
   b. Find the codeword that corresponds to the information word 1101.
   c. Demonstrate that a single error can be detected and corrected.

3. For the convolutional encoder shown in Figure P.1, compute the output coded data when the input data $X = 10111$. Note that the first input bit is the leftmost element of the X row vector.

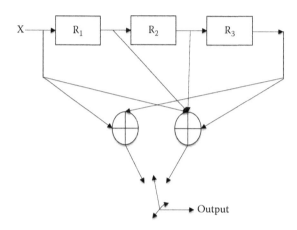

FIGURE P.1
A convolutional coder.

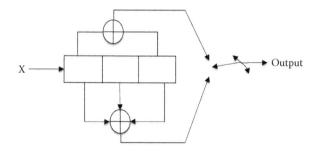

FIGURE P.2
A convolutional coder.

4. Draw a block diagram for a convolutional coder that has a rate of 2/3 and constraint length of 5.

5. A convolutional coder is shown in Figure P.2.

    a. Determine the code rate and constraint length.

    b. Draw the state diagram and code trellis.

    c. Draw the Viterbi trellis.

    d. For the received codeword $X = 11\ 10\ 11\ 11\ 01\ 01$, trace the path sequence.

References

1. Feher, K. (1984). *Digital communications: satellite/earth station engineering*. Englewood Cliffs, NJ: Prentice-Hall.
2. Shannon, C.E. (1948). A mathematical theory of communication. *Bell System Technical Journal, 27*, 379–423.
3. Huffman, D.A. (1952). A method for the construction of minimum redundancy codes. *Proceeding of IRE, 40*, 1098–1101.
4. Fano, R.M. (1963). A heuristic discussion of probabilistic decoding. *IEEE Transactions on Information Theory, 9*, 64–74.
5. Okunev, Iu.B. (1997). *Phase and phase-difference modulation in digital communications*. Boston: Artech House.
6. Lee, E.A., and Messerschmitt, D.G. (1994). *Digital communication*. Norwell, MA: Kluwer Academic Publishers.
7. Lin, S., and Costello, D.J. (2004). *Error control coding: fundamentals and applications*. Englewood Cliffs, NJ: Prentice-Hall.
8. Hagenauer, J. (1996). Iterative decoding of binary block and convolutional codes. *IEEE Transactions on Information Theory, 42*, 429–445.
9. Shamnugam, K.S. (2006). *Digital and analog communication systems*. New York: John Wiley.
10. Peterson, W.W. (1972). *Error correcting codes*. Cambridge, MA: MIT Press.
11. Bose, R.C., and Ray-Chaudhuri, D.K. (1960). On a class of error correcting binary group codes. *Information Control, 3*, 68–79.
12. Pratt, T., and Bostian, C.W. (1986). *Satellite communications*. New York: John Wiley.
13. Barbulescu, S.A., and Pietrobon, S.S. (1999). Turbo codes: a tutorial on a new class of powerful error correction coding schemes. Part II. Decoder design and performance. *Journal of Electrical and Electronics Engineering, Australia, 19*, 143–152.
14. Gallager, R.G. (1963). *Low density parity-check codes*. Cambridge, MA: MIT Press.
15. Sridharan, A., Costello, D., and Tanner, R. (2002). A construction for low density parity check convolutional codes based on quasi-cyclic block codes. *IEEE International Symposium on Information Theory, 481*.
16. Abbasfar, A., Divsalar, D., and Yao, K. (2007). Accumulate repeat accumulate codes. *IEEE Transactions on Communications, 55*(4), 692–702.

17. Hu, X.-Y., Eleftheriou, E., and Arnold, D.-M. (2001). Progressive edge-growth Tanner graphs. *IEEE Globecom Conference*, 2, 995–1001.
18. Richardson, T., and Urbanke, R. (2001). Efficient encoding of low-density parity-check codes. *IEEE Transactions on Information Theory*, 47(2), 638–656.
19. Tanner, R.M. (1981). A recursive approach to low complexity codes. *IEEE Transactions on Information Theory*, 27:5, 533–547.
20. Mackay, D.J.C., and Neal, R.M. (1997). Near Shannon limit performance of low density parity check codes. *IEE Electronics Letters*, 33(6), 457–458.
21. Chung, S.-Y., Forney, D., Richardson, T., and Urbanke, R. (2001). On the design of low-density parity-check codes within 0.0045 dB of the Shannon limit. *IEEE Communication Letters*, 5, 58–60.
22. Kolawole, M.O. (2003). *Radar systems, peak detection and tracking*. Oxford: Elsevier Science.
23. Mitchell, M.T. (1997). *Machine learning*. New York: WCB/McGraw-Hill.
24. Andrews, K., Dolinar, S., Divsalar, D., and Thorpe, J. (2004). *Design of low-density parity-check (LDPC) codes for deep-space applications*. IPN Progress Report 42–159.
25. Viterbi, A.J. (1967). Error bounds for convolutional codes and an asymptotically optimum decoding algorithm. *IEEE Transactions on Information Theory*, 13, 260–269.
26. Viterbi, A.J., and Omura, J.K. (1979). *Principles of digital communication and coding*. New York: McGraw-Hill.
27. Battail, G. (1989). Building long codes by combination of simple ones, thanks to weighted-output decoding. *Proceedings of URSI ISSSE, Erlangen*, 634–637.
28. Berrou, C., and Glavieux, A. (1996). Near optimum error correcting coding and decoding. *IEEE Transactions on Communications*, 44(10), 1261–1271.
29. Hagenauer, J., and Hoeher, P. (1989). A Viterbi algorithm with soft-decision outputs and its applications. *GLOBECOM'89 Dallas, Texas Conference Proceedings*, 1680–1686.
30. Huber, J., and Rueppel, A. (1990). Reliability estimation for symbols detected by trellis decoders (in German). *AEU*, 44, 8–21.
31. Hashimoto, T. (1987). A list-type reduced-constraint generalization of the Viterbi algorithm. *IEEE Transactions on Information Theory*, 26, 540–547.
32. Sundberg, C.W. (1992). Generalizations of the Viterbi algorithm with applications in radio systems. In *Advanced methods for satellite and deep space communications*. Düsseldorf, Germany: Springer-Verlag.
33. Rorabaugh, C.B. (1996). *Error coding cookbook; practical C/C++ routines and recipes for error detection and correction*. New York: McGraw-Hill.
34. Heller, J.A., and Jacobs, I.M. (1981). Viterbi decoding for satellite and space communications. *IEEE Transactions on Communications*, 19, 835–848.
35. Berrou, C., Glavieux, A., and Thitimajshima, P. (1993). Near Shannon limit error-correcting coding and decoding: turbo-codes. *Proceedings of IEEE International Communications Conference*, 2, 1064–1070.
36. Bahl, L., Cocke, J., Jelinek, F., and Raviv, J. (1974). Optimal decoding of linear codes for minimizing symbol error rate. *IEEE Transactions on Information Theory*, 20(2), 284–287.

7

Regulatory Agencies and Procedures

The regulation that covers satellite communication networks occurs on three levels: international, regional, and national. The interaction among the three regulatory levels is briefly discussed in this chapter.

7.1 International Regulations

The reader might query the purpose and need for international regulations. This emerges as the international community realizes that the radiofrequency spectrum and the geostationary satellite orbit are two finite natural resources available to humans. Each of these resources has the unique property of being conserved if it is used properly, and wasted if it is not used properly. To conserve the precious commodity, certain requirements need to be fulfilled:

1. The first essential requirement for the orderly use of the frequency spectrum is the division of the spectrum into separate parts (referred to as bands), where each band can be utilized by one or more communication services.

2. The second essential step is the division of the world into regions. In this regard, the world was divided into three distinct regions: Region 1, Region 2, and Region 3 (see Figure 7.1).

3. The third is the application of preestablished regulatory procedures for the use of frequencies by stations in the same or different service areas in such a way that interference between different services, regions, or nations is avoided.

A regulatory body was formed at a conference in Paris in 1865 called the International Telegraphy Union, subsequently renamed the International Telecommunication Union (ITU), headquartered in Geneva, Switzerland. ITU became the forum wherein nations of the world and the private sector coordinate global telecommunication networks and services, such as the use of the radiofrequency spectrum, and set standards regarding interference levels, equipment characteristics, and communication protocols for all

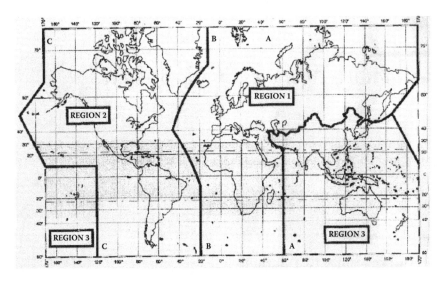

FIGURE 7.1
The three ITU regions of the world. (Courtesy of ITU.)

information and communication technology (ICT) systems—from digital broadcasting to the Internet, and from mobile technologies to 3D TV. ITU is a specialized agency of the United Nations within which governments and the private sector coordinate global telecom networks and services. This forum also establishes procedures for the coordination of frequency spectrum use by member states [1].

With the advent of space technology, new provisions have been adopted in the *Radio Regulations*. These are required to regulate the use of space radiocommunications, particularly the geostationary and nongeostationary orbits.

It should be emphasized that the underlying principle upon which the effectiveness of rules and regulations is based is the voluntary undertaking by each signatory to comply and ensure international cooperation. Though ITU does not have the authority to enforce its rules and regulations, nations have in general adhered to its recommendations as a matter of policy.

To fulfill its mission, ITU established an administrative body, which consists of permanent organizations that convene regional and world conferences to discuss and establish policy to guide its permanent organ and member nations' telecommunication services. The administrative body includes:

1. The General Secretariat

2. Administrative Council

3. The Radiocommunication Sector (ITU-R)

4. The Telecommunication Standardization (ITU-T)

The General Secretariat has an elected secretary general and deputy. The secretariat, being the collective name for the permanent organizations associated with ITU, is charged with discharging the administrative duties of the ITU but has no authority to establish basic policy. The plenary assembly is the main functioning body of the ITU. It usually meets every 4 years, where draft reports and recommendations prepared by study groups within the permanent organs (e.g., ITU-R) are considered and approved. It is during these plenary sessions that the secretary general, deputy, and administrative board members are elected. The plenary assembly also elects the directors of ITU-T and ITU-R.

It should be noted that many of the approved reports and recommendations form the basis for revision of the *Radio Regulations* of the world administrative conferences. Administrative conferences—regional or world—are normally convened to consider specific telecommunication matters. *Radio Regulations* define the rules to be applied in using the spectrum and the orbit, as well as the rights and obligations resulting from this use. The ITU *Radio Regulations* govern satellite communication networks. The approved reports and recommendations are not of themselves obligatory in the same context as the *Radio Regulations*, but serve as standards for use by the world's telecommunication community. World radiocommunication conferences (WRCs) are held every 3 to 4 years, and elect the 12 members of the Radio Regulations Board (RRB), to review and, if necessary, revise the *Radio Regulations*—the international treaty governing the use of the radio-frequency spectrum and the geostationary satellite and nongeostationary satellite orbits. Only member states vote at these conferences, but representatives of other international organizations and private sectors can attend the sessions.

The Telecommunication Standardization (ITU-T) is the primary international body for fostering cooperative standards for telecommunication equipment and systems, covering most interconnected man-made systems, everything from core network functionality and broadband to next-generation services. ITU-T convenes study groups in order to recommend technical specifications for equipment, and to conduct propagation, interference, and other types of analyses.

The work of these sectors (ITU-R and ITU-T) has considerable influence on system planners, network designers, and equipment manufacturers, and to a lesser degree on operators. For this reason, the next subsections briefly describe their origin, responsibilities, and functions.

7.1.1 ITU-R

The ITU-R (formerly CCIR) duties are (as defined under the International Telecommunication Convention, Nairobi, 1982, First Part, Chapter 1, Article II, no. 83) *to study technical and operating questions relating specifically to radiocommunication without limit of frequency range, and to issue recommendations on them.* The objectives of the ITU-R are, in particular:

1. To provide the technical bases for use by the administrative radio conferences and radiocommunication services for efficient utilization of the radiofrequency spectrum and the geostationary satellite orbit, bearing in mind the needs of the various radio services

2. To recommend performance standards for radio systems and technical arrangements, which ensure their effective and compatible interworking in international telecommunications

3. To collect, exchange, analyze, and disseminate technical information resulting from the studies by the ITU-R, and other information available, for the development, planning, and operation of radio systems, including any necessary special measures required to facilitate the use of such information in developing countries

The ITU-R works through the medium of a number of study groups, each dealing with a particular aspect of radiocommunications. The ITU-R has 16 study groups [2]:

BO: Satellite delivery

BR: Recording for production, archival and play-out, film for television

BS: Broadcasting service (sound)

BT: Broadcasting service (television)

F: Fixed service

M: Mobile, radio determination, amateur and related satellite services

P: Radiowave propagation

RA: Radio astronomy

RS: Remote sensing systems

S: Fixed-satellite service

SA: Space applications and meteorology

SF: Frequency sharing and coordination between fixed-satellite and fixed-service systems

SM: Spectrum management

SNG: Satellite news gathering

TF: Time signals and frequency standards emissions

V: Vocabulary and related subjects

The recommendations released by study groups in the Radiocommunication Sector (i.e., ITU-R) constitute a set of international technical standards developed. These recommendations cover [1]:

- The use of a vast range of wireless services, including popular new mobile communication technologies
- The management of the radiofrequency spectrum and satellite orbits
- The efficient use of the radiofrequency spectrum by all radiocommunication services
- Terrestrial and satellite radiocommunication broadcasting
- Radiowave propagation
- Systems and networks for the fixed-satellite service, fixed service, and mobile service
- Space operation, earth exploration satellite, meteorological satellite, and radio astronomy services

The members of these study groups come primarily from the nations most technically advanced in the telecommunication field: these include the United States, Russia, the UK, Canada, France, Japan, Australia, etc. Since the reports and recommendations of these study groups are nonbinding, an effort is made to achieve consensus and unanimity in their findings. Naturally, national interests of the member nations are jealously guided and play in the preparation of the consensus documents.

The most relevant of these groups to satellite communication networks is Study Group S, which deals with fixed services using satellites. A lot of interactions occur between Study Groups S and SF, in defining the criteria for assessing interference between satellite networks and terrestrial terminals in shared bands.

7.1.2 ITU-T

CCITT was formed in 1956 as a result of the merger of two ITU organizations: CCIT (the technical consultative committee for telephone) and CCIF (the technical consultative committee for telegraphy). In 1992, the remodeling of ITU gave it greater flexibility to adapt to an increasingly complex, interactive, and competitive environment, leading to renaming CCITT as ITU-T.

The primary function of ITU-T is to standardize, promote, and define elements in information and communication technologies (ICTs) infrastructure. This goal is accomplished by establishing recommended standards for performance, interconnection, and maintenance of international networks for voice, data, and video communications.

Within the ITU-T structure exist a number of study groups that provide the work classification or specialization for developing the recommendations. The ITU-T has 10 study groups [3]:

II. Operational aspects of service provision and telecommunication management: This group provides service definition, numbering, routing (i.e., identifying mobile devices as they roam from network to network), and management of telecommunication services, networks, and equipment—which were formerly done in Study Group IV.

III. General tariff principles: This group recommends some principles for the harmonization of global interconnection rates that are fair and low without compromising service.

V. Environment·and climate change: This group deals with and publishes guidelines on electromagnetic compatibility and electromagnetic effects. It also initiates studies on methodologies for evaluating the ICT effects on climate change and publishing guidelines for using ICTs in an eco-friendly way.

IX. Television and sound transmission and integrated broadband cable networks: This group carries out studies on the use of telecommunication systems for broadcasting of television and sound programs, and furthermore the use of cable TV (CATV) networks to provide interactive video services and telephone and data services, including Internet access.

XI. Signaling requirements, protocols, and test specifications: This group is responsible for signaling requirements and protocols, including reference signaling architectures and test specifications for existing and emerging networks (e.g., ubiquitous sensor networks [USNs]) enabling seamless mobility between networks. For instance, a signaling requirement may define how telephone calls and other calls such as data calls are handled in the network, for example, signaling system 7 (SS7). More is said about SS7 in Chapter 8.

XII. Performance, quality of service (QoS), and quality of experience (QoE): This group defines end-to-end transmission performance of networks and terminals, as well as the overall acceptability of an application or service.

XIII. Future networks, including mobile and next-generation networks (NGNs): This group leads ITU's work on standards for NGNs and future networks, as well as mobility management and fixed mobile convergence.

XV. Optical transport networks and access network infrastructures: This group develops standards for the backbone architecture, including the key standard for synchronous data transmission over fiber-optic networks, synchronous digital hierarchy (SDH), digital subscriber line (DSL), and on optical access and backbone technologies.

XVI. Multimedia coding, systems, and applications: This group leads all aspects of multimedia standardization, coding (of speech, audio, and video streams), terminals, systems, and applications, including architecture, protocols, security, mobility, interworking, and quality of service.

XVII. ICT security: For example, security standards for electronic authentication over public networks. This group now oversees technical languages and description techniques.

The recommendations have paved the way for standardization of interfaces for public data networks and protocols for international packet-switched data networks, including provision for significant building blocks for integrated services digital networks (ISDNs) and, in turn, integrated services digital satellite networks (ISDSNs). More could be said of their cooperative activities, but we will stop at this point.

7.2 National and Regional Regulations

It is logical to institute a national or regional regulatory body that oversees and manages the country or region's spectrum and adopts national (or regional) legislation that includes, as the basic minimum, the essential provisions of the International Telecommunication Convention. This is necessary because the International Telecommunication Convention and *Radio Regulations* annexed to the convention are intergovernmental treaties, which have been ratified and accepted by the governments of the countries. These governments are bound to apply the provisions of the treaties in their countries and the other geographical areas under their jurisdiction. In reality, a regional body is formed in the context of bilateral agreements with neighboring countries to settle policy or operational issues, for the purpose of coordinating the establishment of telecommunication systems and, importantly, for other items of mutual interest.

To this effect, a country must adopt national legislation that is enforceable, enabling the national regulatory body:

1. To develop regulations for effective use of the spectrum on the basis of national or regional priorities.
2. To coordinate and oversee long-term spectrum management policy and planning.

3. To identify the spectrum requirements that satisfy the need of the country or region.

4. To develop technical standards.

5. To monitor, detect, and resolve operational irregularities, harmful interference, and technical ambiguities.

6. To conduct negotiations relating to frequency spectrum management and related matters that may improve technical and administrative cooperation with other countries and international organizations.

7. To keep accurate and up-to-date data records. This is a requirement of utmost importance for effective national, regional, and international coordination; licensing and enforcement activities; policy formulation; interference investigation and resolution; and financial considerations.

8. To participate in ITU study groups where national or regional interests are safeguarded. An example is CSIRO—the national scientific, industrial, and research organization—which represents Australia in ITU meetings, particularly in propagation, spectrum issues, and other areas of radio system development.

9. To provide training to personnel for frequency management including seminars organized by the ITU-R and ITU-T, exchange of staff with other countries, and hands-on operational experience on the job.

The national or regional spectrum management body should therefore administer day-to-day spectrum applications originating from the public and governmental sectors. In some instances, the flow of activities concerned with the approval, licensing, spectrum allocation, and international coordination of commercial and military satellite systems may be handled by different sections of the national or regional body according to the requirements and resources of the country concerned.

7.3 Summary

The regulation that covers satellite networks occurs on three levels: international, regional, and national. The underlying principle on which the effectiveness of the international and regional regulations is based is the voluntary undertaking by the governments that signed the regulations to comply with them and ensure international and regional cooperation. The regulatory bodies formed by these governments have the legislative power, within their jurisdiction, to ensure and enforce an orderly distribution of frequency spectrum in compliance with the treaties acceded to and ensuring no interference in the spectrum assigned to the operators. The underlying principle and the interaction among the three regulatory levels have been discussed in this chapter.

Problems

Study the radiocommunication regulatory agency of your country. Is there any way you can rationalize the various departments within the agency?

References

1. ITU. (2012). A brief history of ITU. Geneva. http://www.itu.int/ (accessed September 1, 2012).
2. Radiocommunication. http://www.itu.int/ITU-R/ (accessed September 10, 2012).
3. Standardization. http://www.itu.int/ITU-T/ (accessed September 10, 2012).

8

Mobile Satellite System Services

8.1 Overview

Until the introduction of the International Maritime Satellite Organization (INMARSAT) to serve ships at sea, all satellite communication links were between fixed locations. The introduction of maritime services served to focus attention on the whole field of mobile communications and opened up a range of novel opportunities. These include:

1. Standardization of system solutions. The most successful of standardized systems is *Global System for Mobile Communications* (GSM). The GSM architecture is based on a terrestrial switched network.

2. Provision of personalized services and mobility as well as interactive telecommunication systems (e.g., short text messaging) using small handheld terminals and the need for a single telephone number worldwide for a roaming traveler.

3. The need to expand or enhance the capability of the existing terrestrial and cellular communication networks to sparsely populated regions, remote areas, or areas where a communication infrastructure is suddenly destroyed or has never existed.

4. Provision of very accurate navigational data, ship-to-shore search, and rescue messages in emergencies.

5. Direct broadcast television, instant broadcast of impending danger, and specialized business services.

6. Systems tracking for logistic reasons.

This chapter examines the basic structure of a mobile satellite system (MSS) and its interaction with land-based public switched digital networks, in particular the *integrated services digital network* (ISDN). Since the services covered by ISDN are also, in principle, provided by a mobile satellite system network, it is considered necessary to discuss in some detail the basic architecture of ISDN as well as its principal functional groups in terms of reference configurations, applications, and protocol. The chapter finishes by briefly examining the cellular mobile system, including cell assignment and the internetworking principle, as well as technological shortcomings to providing efficient Internet access over satellite links.

8.2 Mobile Satellite Systems Architecture

A *mobile satellite system* (MSS) is a system that provides radiocommunication services between:

1. Mobile earth stations and one or more satellite stations
2. Mobile earth stations by means of one or more satellites
3. Satellites

Figure 8.1 shows the basic architecture of an MSS with a land-based digital switched network (LDSN) and intersatellite cross-link. Assuming that a new-generation mobile satellite is being designed for Figure 8.1, the total spacecraft system, such as power, guidance and control, and data handling, would have advanced technology components. The satellite would contain onboard digital signal processors (DSPs) and memory for onboard data processing capability, and onboard fast packet switches. The onboard fast packet switches would be capable of supporting space-optimized traffic from multiple earth stations.

The DSP will be responsible for resource management and control, including encryption/decryption, channelization, demodulation, and decoding/encoding. This functionality has been discussed in previous chapters. As stated earlier, since most, if not all, of the services covered by the MSS are, in principle, provided by terrestrial switched digital networks (e.g., ISDN), we will attempt to explain some of the concepts applicable to ISDN that have not been dealt with in previous chapters.

The network routers (gateways) allow:

Seamless intersatellite cross-link; that is, direct data transfers from one satellite to another.

Seamless connectivity for users anywhere in the world through mobile/fixed earth stations and public switched digital networks.

Traffic shaping, resource accounting, cache for traffic redirection and load sharing, and integrated network management to support myriad simultaneous connections per satellite.

Any mobile station registered on the mobile satellite network is interconnected to any available channel of the network interface gateway (NIG)

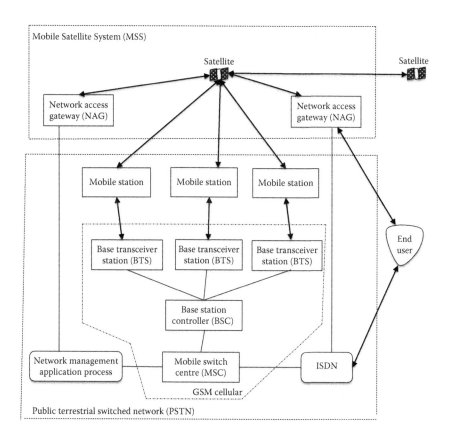

FIGURE 8.1
Block diagram of a mobile system with terrestrial switched network and intersatellite cross-link.

through proper channel assignments issued by the network access gateway (NAG).

When the satellite illuminates a particular area or region, the mobile satellite system routes intended messages (e.g., telephone calls, data, etc.) through the ground networks (e.g., ISDN), ground stations, or directly to the user terminals. User terminals can be personal terminals for individual subscribers or multiuser terminals for corporate (e.g., Internet providers, communication resellers, etc.) and communal residential.

The public terrestrial switched networks, called in this text *land-based digital switched network* (LDSN), contain the integrated services digital network (ISDN) and mobile communication systems to provide end users with efficient communication services between fixed and fixed terminals, fixed and mobile terminals, and mobile and mobile terminals. In the network arrangement shown in Figure 8.1, any mobile station using the services of LDSN can communicate both signaling and bearer traffic to the base transceiver station (BTS) that provides the most favorable radiofrequency (RF) signal. This establishes an association between the mobile station's geographic location and the closest BTS. As the mobile station moves from the coverage area of one BTS to another, the first association is released and a new one is formed. This procedure is called *handover*. The base station controller (BSC) and mobile switching center (MSC) manage radio resources, channel assignments, and handover services. A single BSC can control multiple BTSs. A single MSC can control multiple BSCs. Multiple MSCs may reside within a single LDSN. The network *management application process* (MAP) defines services for signaling among several MSCs. In principle, all the services MAP defines and provides are applicable to the MSS. This duality of services becomes obvious when considering MAP in Section 8.2.1 since the services are offspring of the parent class of LDSN.

As previously discussed in Chapter 5, sufficient access can be provided by satellites; however, the type and number of accesses provided depends on the access scheme employed. As noted in Chapter 2, satellite-based transmitter and receiver chains can operate through high-gain spot beam antennas deployed in arrays of multiple beams to cover the desired service area on earth. If an antenna beam is focused at a fixed area, we can use a single satellite network access gateway to manage and control all mobile stations located within the spot beam to provide direct data link access. Multiple network interface gateways may be deployed, in circumstances where service areas are sparsely distributed, to provide local terrestrial interconnection between the satellite and public terrestrial networks for efficient routing of circuit-switched bearer services and teleservices.

At this stage, it is considered necessary to discuss in some detail the basic architecture of ISDN as well as its principal functional groups in terms of reference configurations, applications, and protocol.

8.2.1 Integrated Services Digital Network (ISDN)

Integrated services digital network (ISDN) is a telecommunications network that provides end-to-end digital connections to end users and other network facilities. An MSS network should also, in principle, provide the services covered by ISDN. However, network standards of MSS networks should complement those of ISDN.

The principles of ISDN recommended by ITU-T, particularly Group XI (see Chapter 7), govern its development, protocols, and standards. Recently, ITU-R Study Group S (see Chapter 7) became involved with digital baseband processing techniques, particularly for fixed and broadcasting satellite services. As such, their work has an impact on the standardization of integrated digital services that would be provided by a satellite digital network [1].

The next subsections discuss structural aspects of the ISDN's communication capabilities because they [2]:

1. Impose a logical structure on the communication-related aspects of a functional group
2. Provide a convenient framework within which we may define the physical and logical relationship between communicating functional groups, the types of information flows they exchange with each other, and the protocols that govern their interaction
3. Describe the characteristics of the services that an ISDN makes available to the end users

ISDN forms the backbone of terrestrial digital networks. To a considerable extent, ISDN topology depends on the nature of the users and network providers. A high-level functional representation of an ISDN is shown in Figure 8.2.

With this high-level functional configuration, we are able to partition the network into a set of components that interact with each other at well-defined interfaces. By this partition we are able to specify each component function individually and define the relationship between the end user and those of the network.

A number of books and papers have been devoted to this topic alone, including those listed in the reference section [3–8]. As a result, a brief description of principal functional groups of ISDN is given in terms of reference configurations, that is, abstract representations of its overall structure into a number of functional groups.

Central to an ISDN is the interexchange network (IEN), which consists of the physical and logic components of the backbone transmission network, including switching facilities. These facilities comprise a number of switching exchanges (called digital transit exchanges) and a series of transmission media that interconnect the exchanges in a distributed network topology.

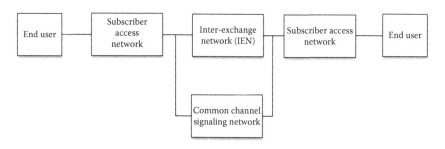

FIGURE 8.2
ISDN configuration.

The exchanges provide the physical terminations of the interexchange transmission media and select and allocate transmission and storage facilities for the end user connections. Since the end user connections can be circuit switched or packet switched, both switching exchanges are normally present in an ISDN.

Packet switching is a method of transmission in which small blocks of data called *packets* traverse in a store-and-forward method from source to destination through the intermediate modes of the communication network and in which the resources of the networks are shared among many users. Circuit switching is a method of transferring information in which switching and transmission functions are achieved by permanently allocating a number of channels or bandwidth between the connections.

The common channel signaling network (CCSN) provides the physical and transmission capacity for the transfer of connection control signals between components of IEN. In essence, it provides the functions required for the control, management, and maintenance of ISDN.

The subscriber access network (SAN) consists of the part of ISDN between the end user or subscriber and CCSN and IEN. SAN can further be decomposed into several reference points, ensuring physical or virtual interfaces between the functional groups. A further decomposition of each functional group is omitted in this text. However, an inquiring reader should consult Helgert [2], Ronayne [9], and Wu [10] to start.

The functions of each of these interconnected components (IEN, CCSN, and SAN) do not necessarily correspond in a one-to-one manner when specifying physical parts. The functions of each of these parts may be contained in one or more functional groups that interact with each other across reference points. For example, individual functions may be implemented in one or several pieces of equipment, and one piece of equipment may perform functions in more than one group.

In order to balance the requirements of efficiency, universality, flexibility, minimum complexity, and low cost on the one hand, and the wide variety of existing and evolving devices and applications on the other, three distinct end user-to-network interfaces have been identified: *basic access, primary access,* and *broadband access,* as shown in Figure 8.3. These three interfaces vary practically in the level of physical and logical complexity of the functional groups, the degree of end user control over the network connections, and the types of connections required.

In terms of data rates, the basic access operates up to 100 kbit/s; it is designed primarily for single or multiple low-speed terminals, digital telephones, and residential and small business users. The data rates for primary access are up to 2 Mbit/s and are intended for slow-scan, high-quality audio devices, high-speed facsimile, local area networks (LANs), and digital private automatic branch exchanges (PABXs). The data rates of the broadband access type are in the several Gbit/s range. The broadband access type of user-network interfaces is designed for video terminals,

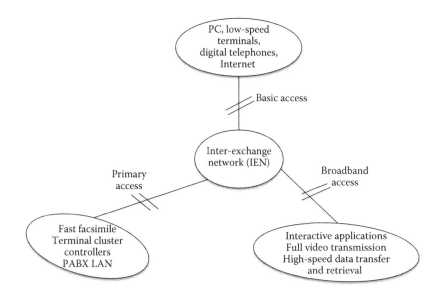

FIGURE 8.3
Types of user-to-network interfaces and ISDN applications.

multimedia multiplexers, high-speed file transfer, and document retrieval, to name a few.

8.2.1.1 Channel Types and Structures

The three aforementioned types of user-to-network interfaces, namely, basic access, primary access, and broadband access, are also characterized by their transmission capacity in terms of a set of component channels, which can also be aggregated into synchronous and asynchronous channel structures. Channel structures refer to the combinations of different types of channels. A channel is a specific portion of the information carrying capacity of a network interface and is specified by a specific transmission rate.

The transmission rate of a component channel in a synchronous channel structure is constant but may vary from one component to another, whereas in the asynchronous channel structure, a component channel is dynamically created by the instantaneous transmission requirements of the application. An asynchronous channel structure has a feature akin to "bandwidth-on-demand," allowing a theoretically unlimited variety of component channels.

8.2.1.1.1 Channel Types

Depending on the usage and information carrying mechanism, ITU-T [11] classified channels into three types: B, D, and H. As the needs and technologies change, the rates specified in this section for each channel may change.

The B-channel operates at a synchronous data rate of 64 kbit/s in full duplex mode, that is, carrying either voice or data in digital form. B-channels can be used at separate user-to-network interfaces to provide access to a variety of communication modes within the ISDN. For example, two B-channels may be attached to each other via network-provided circuit-switched or packet-switched connections, or by permanent or temporary connections based on circuit switching or packet switching. The B-channel can carry layered protocols according to Recommendation X.25. X.25 defines how connections between user devices and network devices are established and maintained. X.25 allows many options for each network to choose when deciding which features to support and how certain operations are performed.

The D-channel primary function is to carry signaling information for the control of circuit-switched connections involving one or more B-channels between the end user and the network. Two types of D-channels have been defined: layered protocols and X.25. Both operate at synchronous data rates of either 16 or 64 kbit/s in full duplex. In addition to signaling information for circuit switching, a D-channel may also be used to carry teleaction information and packet-switched data. The term *teleaction* means "acting at a distance," where a physically remote or geographically distributed audience is collaboratively controlling a remote agent.

The H-channel is designed for a variety of user information functions. A distinctive characteristic is that an H-channel does not carry signaling information for circuit switching by the ISDN. Currently, there are two types of H-channel: H0, and H1 with successively higher rates (see Table 8.1). Channel H1 exists in two versions: H11 and H12.

The H0-channel operates at a full duplex synchronous data rate, six times the B-channel rate, which is a submultiple of ITU-T multiplex structures (e.g., 1.544 and 2.048 Mbit/s). The H0-channel is suitable for standard broadcast digital audio programs.

The H1-channels are designed to carry very high speed facsimile signals, standard compressed videoconferencing signals with reduced spatial resolution and movement, and transmission capacity for private network trunks.

TABLE 8.1

H-Channel Transmission (Data) Rates

Component Channel		Data Rate kbit/s	Multiple of B-Channel Rate
	H0	384	6
H1	H11	1536	24
	H12	1920	30

For circuit-switched connections, none of the H-channels carries signaling information.

8.2.1.1.2 Channel Structures

Channel structures define the maximum digital carrying capacity in terms of bit rates across the ISDN interface. The bit rates and channel combinations are chosen to match the aggregate data rates of the intended applications. The fundamental objective of channel combinations is to create the possibility of several simultaneous and independent voice or data calls, by either the same or different terminals over a single user-to-network interface, particularly for synchronous channel structures, discussed as follows.

The basic channel structure, corresponding to *basic access*, is comprised of two B-channels and one D-channel. As an illustration, for an aggregate data rate of 144 kbit/s, we can choose

$$2B + 1D = (2*64 + 16) \text{ kbit/s} \tag{8.1}$$

The primary channel structure corresponds to *primary access*, designed to accommodate applications requiring more than two simultaneous connections or data rates in excess of 64 kbit/s. For example, for an aggregate data rate of 1.544 Mbit/s, we choose either

$$n\text{H0} + m\text{B} + \text{D} \tag{8.2a}$$

where the integers n and m have values in the range

$$0 \leq n \leq 3$$
$$0 \leq m \leq 23 \tag{8.2b}$$
$$6n + m \leq 23$$

or

$$n\text{H0} + m\text{B} \tag{8.3a}$$

where

$$0 \leq n \leq 4$$
$$0 \leq m \leq 24 \tag{8.3b}$$
$$6n + m \leq 24$$

Justification for using either (8.2) or (8.3) depends on whether any call control information signals are transmitted. As the aggregate data rate increases, the channel combinatorial aspects vary with intended applications.

Naturally the broadband channel structure corresponds to *broadband access*, appropriate for multimedia applications, which include data, full video, interactive applications, and call control signals. As an illustration, for the transmission of 129 Mbit/s, we can choose several combinations for several applications, such as

$$4H12 + 2B + D \tag{8.4}$$

where functions can be distributed across the channels, for example:

The four H12 channels to provide capacity to convey fast data transfer, etc.

The remainder of the channels, that is, (2B + D), are to provide support for low-data rate information.

For asynchronous channel structures, information flows are not constant. Figure 8.4 shows a basic asynchronous channel structure. The design of asynchronous channel structures is based on the concept of the transfer cell. A *cell* is an information frame containing a fixed and integral number of octets of information. The transfer cell provides a basic unit of capacity. A particular transfer cell carries within it a header for channel identification, and the other portion of the cell is used for carrying the useful information.

If an asynchronous channel uses an average of one out of every K cells, its average data rate $\langle r_b \rangle$ can be expressed as

$$\langle r_b \rangle = \frac{R_r(L_a - h)}{KL_a} \text{ bit/s} \tag{8.5}$$

and the average time between successive channels is

$$T_c = 8\frac{KL_a}{R_r} \text{ s} \tag{8.6}$$

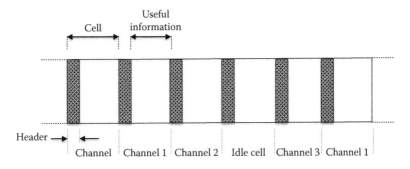

FIGURE 8.4
An asynchronous channel structure.

where K = number of cells, L_a = number of octets in a cell, R_r = transmission rate (bit/s), and h = number of octets in the header.

The total capacity of the channel structure must be sufficient to at least accommodate the application with the highest data rate that must be supported over the user-to-network interface.

Unlike the synchronous channel structures with a fixed channel arrangement and known capacity, the component channel of the asynchronous channel structures dynamically provides a data rate matching the requirements of the applications. This arrangement offers flexibility in efficiently supporting a wide and dynamically varying spectrum of applications with different data rates and usage patterns.

In general, the channel structures provide the transmission capacity for the exchange of information between functional groups on the two sides of the user-to-network interface. The principle of functional layering inherent in the *open systems interconnection* (OSI) leads to a more general representation of the total communication capacity of a functional group. Total communications of a functional group mean, for example, the interchange of information between the user and the network elements, between network elements of the same network, and between different networks. Since the OSI reference model was designed to allow various "open" systems to communicate, more will be said later on the OSI reference mode. A system that communicates with other systems is defined as being *open*. As such, an open system is standards (protocols) based instead of propriety based. The ISDN protocol reference model is the basis for a set of ISDN protocol configurations. There is a functional relationship between OSI and ISDN; this relationship will become obvious to the reader when the two protocol models are discussed.

8.2.1.2 Error Performance Standard

The error performance standard P_{EFS} for digital transmission was defined by Mahoney [12] as

$$P_{EFS} = (1 - P_e)^{n_b} \tag{8.7}$$

where P_e is the bit error rate (BER) defined by Equation (3.12) in Chapter 3. Expression (8.7), for some time, was adopted by ITU-R and ITU-T study groups as the performance measure for digital transmission systems, including ISDN [13–15]. Further study groups' studies have been conducted, and are still continuing, leading to a number of recommendations, including G.821, G.826, G.828, G.829, G.8201, I.356, and the M.21xx series. This error performance expression will be valid for digital satellite communication purposes if the transmission channel error statistics are known, for instance, the burst error information, and data speed in both synchronous and asynchronous mediums and over the shared channel.

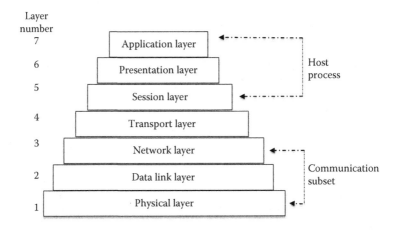

FIGURE 8.5
The OSI reference model.

8.2.1.3 OSI Reference Model

The OSI model is a layered architecture, divided by the International Organization for Standardization (ISO) into seven layers; see Figure 8.5. Each layer is a module performing one primary function and has its own format of communication instructions or protocols. The protocols used to communicate between functions within the same layer are called *peer protocols*. A protocol is a set of necessary procedures or a format that establishes the timing, instructs the processors, and recognizes the messages. The next few paragraphs summarize the seven functions performed by the layers.

The OSI layers are divided into two important subsets: the *host process* (or customer portion of the system) and the *communication subnet* (also called *communication subnetwork* or the carrier portion of the system). The host process is comprised of the upper three layers (the session, presentation, and application layers). The middle layer, the transport layer, is the first end-to-end layer, which acts as a buffer between the two subsets. In some literature, the transport layer is grouped with the host process.

The lowest layer of the OSI reference model is the *physical layer*. It handles the data transmission between one node (e.g., host workstation) and the next, monitors data error rates, and defines anything generally associated with the physical transmission of the bit stream.

The *data link layer* maintains a reliable communication link between adjacent nodes. Once communication is established between two nodes, their data link layers are connected physically through the physical layer and logically through peer protocols. Data transmissions through this layer are synchronized so that frames (or packets) are sent in the same sequence as received.

The *network layer* creates the logical paths, known as virtual circuits, along the communication subnet for transmitting data from node to node. Only the network layer knows the virtual circuits. Switching, routing, and controlling the congestion of information frames within the subnet are functions of this layer.

The *transport layer* manages the successful node-to-node transportation of data. This layer handles packet errors, packet order, and other critical transport issues. The major difference between the data link and transport layers is that the data link domain lies between adjacent nodes, whereas that of the transport layer extends from the source node to the destination (or end to end) within the communication subnet.

The transmission link between two nodes is maintained by the *session layer*, ensuring that point-to-point transmissions are established and remain uninterrupted. Once the communication session is finished, this layer "performs a disconnect" between the nodes.

The *presentation layer* formats and encrypts data to be transmitted across a network. It provides a data manipulation function, not a communication function.

The *application layer* provides application services for end users. These services include file transfers, file management, e-mail, virtual terminal emulation, remote database access, and other network software services.

In summary, the OSI reference model of network communications provides, in particular, a basis for nodes to communicate within the same network and for networks to communicate with other networks using different architectures.

8.2.1.4 ISDN Protocol Reference Model

As previously defined, a protocol is a set of necessary procedures or a format that establishes the timing, instructs the processors, and recognizes the messages. A protocol reference model is a framework for the hierarchical structuring of functions in a system and its interaction with another system. The ISDN protocol reference model is based on the layered structures and signaling system 7 (SS7) for both packet switching and circuit switching with type B- and D-channels, already discussed in this chapter.

Conceptually, the ISDN protocol reference model may be constructed by taking into consideration three planes of communication in a particular network: the user messages (including digitized voice, data, etc.), control information (such as access control; network usage control including *in-band signaling*, e.g., X.21 and X.25, and *out-band signaling*, e.g., SS7), and local station management aspects. The basic protocol building blocks of the three planes of communication are shown in Figure 8.6. In the figure, the user message plane is denoted by U, the control information plane by S, and the management plane by M.

The primary function of the U-plane is to convey user information flows, user-to-user signaling flows, and user management information flows.

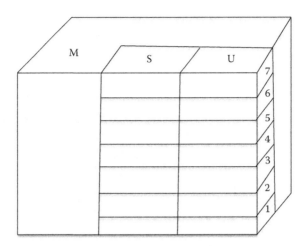

FIGURE 8.6
A protocol reference model building block: U, message; S, control; and M, management.

In terms of channel structures, the U-plane is associated with the B- or H-channels but may also involve the D-channels. The U-plane is, in general, incapable of changing the state of the connection itself.

The main function of the S-plane is to control information flow as well as the management of information flows between a user and a network management facility. The S-plane is normally associated with the D-channels. Both the U- and S-planes are further decomposed into seven layers: numbered from U-1 to U-7 and S-1 to S-7, respectively. These layers' flows are conveyed in a logical sense between peer entities in the appropriate plane of communicating functional groups.

Physically all the layers' subflows are transmitted as a single flow over the physical transmission medium that interconnects the functional groups. To identify which layer belongs to what, names such as *network layer* or *data link layer* are used. The functions of each of these layers may be contained in one or more functional groups that interact with each other across the user-to-network interface for the purpose of transferring information flows.

Most interactions between U and S layers go through the management plane, M, the exceptions being in the higher layers. The main function of the M-plane is to manage the U and S resources, facilitate prompt exchanges of information flows between them, and convey management information to another system, the management application process (MAP).

The preceding discussion allows us to say that a logical transfer of the subflows (i.e., at the layers level) between functional groups exists and is governed by peer-to-peer protocols between pairs of U-plane or S-plane entities. Additionally, the protocol of each layer is independent of the protocol of the

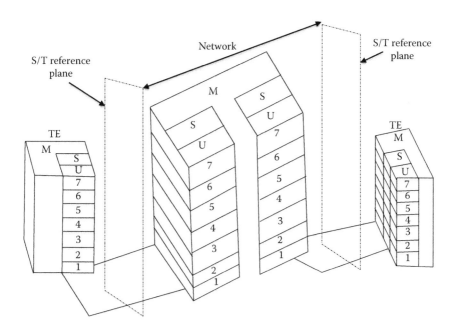

FIGURE 8.7
An ISDN protocol reference model.

other layer. Consequently, a general ISDN protocol reference model can be developed with the aid of Figure 8.7, as follows:

For the reference model within the network segment, different media connections may be used between two inner building blocks.

The U-plane can be represented by the three lower layers for network elements that perform information exchange and transfer functions.

The seven layers are used for network elements that perform end system functions.

The notation TE in Figure 8.7 denotes *terminal* (or station) *equipment*, a functional group that includes functions necessary for protocol handling, interfacing, maintenance, and connection to other equipment. S/T is the *signaling point* or user-network point whose primary function is to *transfer* signaling messages from one signaling link to another. The other codes are as previously defined. The various information flows exchanged between peer planes and peer entities in different functional groups are carried over the same physical medium. The physical and logical interface is the connection point to external networks. ISDN incorporates the physical, data link, network, and transport layers of the OSI model. It should be noted that the information flows could be carried over separate component channels of the existing channel structure or be multiplexed onto a single-component channel.

Advances in technology demand further expansion in the development of protocols covering full multimedia services (broadband and Internet). As such, there is a need for protocols that:

Define a universal user-access interface for broadband satellite multimedia services that (1) is open to different satellite system implementations and (2) provides both the *Internet Protocol* (IP) and real-time seamless broadband service provision

Converge both fixed and mobile multimedia networks toward a global multimedia mobility architecture, that is, by merging service functions derived from the *universal mobile telecommunication system* (UMTS) and the new-generation mobile services (simply called IMT-2000, mobile multimedia phones or smart mobiles) and from fixed broadband access

Separate higher-layer access network functions, which traditionally hide the lower-layer-dependent functions from the end user and the core network

Demonstrate how to apply higher-layer access network functions or layer-dependent functions to other existing satellite access network functions and planned systems

Allow measurement of *quality of service* (QoS) and flexibility for optimized or proprietary air interfaces to satellite systems.

8.2.2 Cellular Mobile Systems

An example of cellular mobile systems is the Global System for Mobile Communications (GSM), one of the most successful standardized systems. GSM architecture is based on public land mobile network (PLMN), a segment of land-based digital switched network (LDSN), shown in Figure 8.1. Some of the attributes of mobile communication systems include [16]:

The extensive frequency reuse with a large number of widely dispersed low-power transmitters located at the base stations

Computer-controlled capabilities to set up automatic handovers from base station to base station when the signal strength or transmission distance can be improved to a more acceptable value

The major difference between mobile satellite communications and land mobile communications includes path loss, noise environment, and fading characteristics. Path loss and noise have been discussed in Chapters 3 and 4, and fading is covered in Section 8.2.2.4. The functional components comprising mobile communications have been described in Section 8.2. Our next task is to describe the basic call setup for a mobile (cellular) terminal.

8.2.2.1 Call Setup

When a call is placed, the handset first senses the presence or absence of cellular frequencies and attempts to place a call through the local cellular networks if they are available. If cellular service is blocked or not available, the network operator routes the call through the satellite for an acknowledgment of the call status. User circuits established through a particular base transceiver station enter the ground networks (e.g., ISDN) through the satellite gateway located within the region being served. Call setup is thus established through an order wire to a mobile switch center (MSC). Often a separate order wire is assigned to each satellite antenna beam. To make a circuit request, the user terminal (handset) transmits requests via order wires to the satellite. The user (1) is assigned a beam that provides the strongest signal to the base transceiver station and (2) is then instructed to use a suitable power level and particular access technique appropriate to the beam. In some cases with very long callers, circuit transfer to another satellite is performed without the callers' participation.

Where the user is moving outside the beam coverage, the call is handed over to the next beam—a process called *handover*. For call setup via order wire, each base station constantly monitors the strength of the modulated signals it is receiving from each user. When the quality of this signal falls below certain preassigned norms (values) because the direct path signal may have been attenuated or the cellular user is moving outside the beam coverage, the base station sends a request to the centralized cellular switch (MSC) asking it to attempt a handover. The base station is usually located in the barocenter (center of the cell) of the service area. The switch then automatically asks each adjoining base station to scan the frequency being used and report the quality of the signal it is picking up. If substantial improvements can be made, the switch automatically orders a handover.

8.2.2.2 Cell Size and Division

8.2.2.2.1 Cell Size

When planning a cellular system, the main objective is to achieve the maximum use of available radio spectrum, besides ensuring low interference and quality of service. It is possible to group geographic areas of mobile users into cells. The size of the cell depends on the population density of the users. Cells are smaller where the most users are expected. The shape of a cell is made such that interlocking cells obtain real coverage. On paper, the shape that allows interlocking to occur is any member of the family of polygons

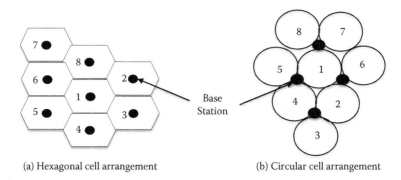

(a) Hexagonal cell arrangement   (b) Circular cell arrangement

FIGURE 8.8
Cell arrangements for geographical service areas.

(see Figure 8.8a). In real life, however, the shape is determined by the radiation pattern of the transmitter's antenna. This antenna may be omnidirectional in rural areas with sparse populations. The complete coverage is known as *cellular*, thereby providing the term *cellular radio* [17, 18].

Three factors affect the number of channels that can be used in a particular area [19–21]:

1. The frequency spectrum available
2. The minimum size of the cells that can be achieved without endangering lives due to excessive radiation
3. The percentage reduction in the quality of the link that can be tolerated due to co-channel interference

It is possible to formulate a method of obtaining a realistic estimate of the number of cells. Let's define the following items for the purpose of developing an algorithm:

Geographic area being considered = A_g (km²)

Average radius of a radio cell = r_c (km)

Channel gain (due to dynamic assignment or trunking technique) = ζ

Cell cluster size = N_c

Total bandwidth assigned to the service = Δf_{ch} (Hz)

Bandwidth allocated to each user = Δf_u (Hz)

For a uniform cell arrangement, the number of cells in a service area may be written as

$$n_c = \frac{A_g}{\pi r_c^2} \qquad\qquad (8.8a)$$

for a circular cell's arrangement, as in Figure 8.8(b). For an arrangement based on any member of the polygon family, the number of cells in any polygonal service area can be expressed as

$$n_c = \frac{A_g}{A_p} \tag{8.8b}$$

where A_p is the generalized area of the polygon (as in Figure 8.8(a)), where all of its sides and angles are equal, expressed as

$$A_p = nr_c^2 \tan \tfrac{\alpha}{2}$$

$$= \frac{nar_c}{2}$$

$$= \frac{1}{4}na^2 \cot \tfrac{\alpha}{2} \tag{8.8c}$$

where, using Figure 8.9 as a guide, a = common length, n = number of sides, and

$$\alpha = 360°/n \tag{8.8d}$$

The number of radio channels in each cell is

$$n_{ch} = \frac{\Delta f_{ch}}{N_c \Delta f_u} \tag{8.9}$$

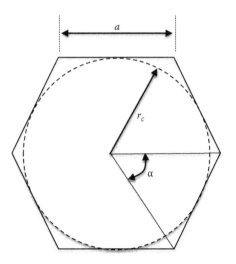

FIGURE 8.9
Geometry of a polygon.

The total number of users that can be supported is

$$N_{users} = n_c \zeta n_{ch} \tag{8.10}$$

The above expression shows that the success of a cellular arrangement depends on the spectrum allocated per user—the term Δf_u—rather than channel bandwidth assigned.

8.2.2.2.2 Cell Division

Finding a solution to limited radiofrequencies, alleviating problems of congestion, and providing adequate services are not easy. Since telephone service providers are in the business of making a profit, the issue of restricting the number of subscribers and an imposition of aggressive tariffs on call durations are the last options because service delivery will be substandard. Subscribers penalize such telephone providers. A realistic option is to improve the economy of networks by applying more sophisticated network-planning methods. Cell division and smaller cell techniques help to increase system capacity because they enable a more efficient reuse of frequencies within a short distance.

Traditionally, the first phase in cellular network building is a larger cell network that rapidly provides the coverage desired. A larger cell network uses high transmitter powers in the base station. Both directional and omni-directional antennas can be deployed according to the required coverage and frequency reuse demands. When the traffic increases, larger cells can further be divided into smaller ones. When reducing cell size, we can use the following techniques:

1. Add new cells by building smaller cells inside the larger one. When a cell is split up into smaller ones, omnidirectional base stations are replaced by more stations with a larger antenna height and reduced transmitting power. The smaller the cell, the lower the required transmitting power. We can calculate the required transmitting power from the required field strength at the edge of the estimated cell coverage area; see Figure 8.10.

2. Split cells into sectors (also called sectoring) by using directional antennas. When frequency economy is not a limiting factor, adding radio channels in the omnidirectional cell can increase the capacity of the network. In dense networks where the same frequency must be reused several times, sectoring gives better capacity since more frequencies can be allocated to one cell and small-cell clusters can be used (see Figure 8.11).

3. Overlay cells (as in Figure 8.11). In high-traffic areas it is difficult to achieve full coverage due to nonuniformly distributed traffic and small cell sizes. A way of overcoming blank spots is to use the

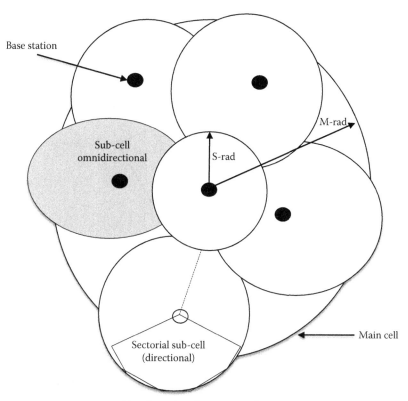

M-rad = main cell coverage radius
S-rad = sub-cell coverage radius

FIGURE 8.10
Cell division.

umbrella cell technique. Umbrella cells are large cells that cover one or several cells and the areas between them. The basic idea behind this technique is that channels are used only when no small-cell channels are available. In other words, the calls are routed to small cells. If none of the small cells cover the spot where a user is trying to make or receive a call, one of the channels from the overlapping umbrella cell is allocated to the cell.

Keep in mind that topology and other geographical factors set limits to network planning. In practice, network planning is a process of experimenting and iterating that seldom, if ever, ends up with a regular cluster composition; the product is typically a group of directional cells of varying sizes turned in different directions. There are technical limitations as well, notably handover and interference. Handover limits the size of the cell when a call is switched from a fading traffic channel to a stronger

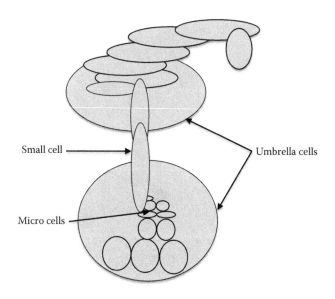

FIGURE 8.11
Splitting cells into smaller cells.

one. If the cells are so small that the cellular phone has to hand over to another base station several times during a call, the quality of speech cannot be the best possible. At the same time, the infrastructure of the network is heavily loaded. In high-density system traffic, channels easily interfere with each other, resulting in adjacent and co-channel interference. Adjacent and co-channel interference can be avoided by careful frequency planning.

8.2.2.3 Frequency Reuse

As the assigned frequency band Δf_{ch} is totally used by the cluster of N base stations, it must be repeatedly reused if contiguous radio coverage is to be provided. This means that N-cell clusters must be tessellated, as shown in Figure 8.12.

The effect of tessellating clusters is that a roving mobile station traveling in, say, cell 1 of a particular cluster will experience interference from the base station located in cell 1 of the other clusters, which may be transmitting to mobile stations in their cells using the same frequency. This is *co-channel interference*. To deliver an acceptable speech quality to mobile users, with minimum co-channel interference, the reuse distance d_i must be at least

$$d_i = r_c \sqrt{3N} \qquad (8.11)$$

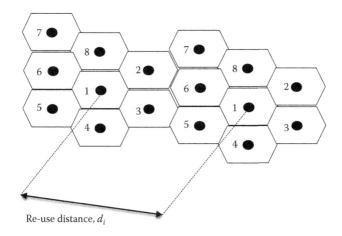

FIGURE 8.12
Tessellated clusters of cells, five cells per cluster.

If the same bandwidth Δf_{ch}, but not the same frequency band, is assigned to each cell and each user is allocated a bandwidth Δf_u, then the maximum number of channels per cell is

$$n_{c\,\text{max}} = \frac{\Delta f_{ch}}{\Delta f_u} \tag{8.12}$$

If each cell covers an area A_c, then the channel's density may be expressed as

$$d_c = \frac{\Delta f_{ch}}{A_c \Delta f_u} \tag{8.13}$$

8.2.2.4 Fading and Diversity Schemes

In addition to changes in polarization due to the Faraday rotation effect in the ionosphere, mobile satellite communications have crossover from polarization to polarization due to the structure of the mobile services. This rotational effect causes signal fading. Of course, the crossover effect precludes the use of two different orthogonally polarized waves to increase channel capacity.

A diversity scheme is needed to combat fading. Several diversity schemes have been employed, including space diversity, frequency diversity, micro- and macrodiversity, and time diversity. The mode of operation in which two or more base stations serve the same area is called *macrodiversity*. *Microdiversity* refers to the condition in which two or more signals are received at one base site or mobile site. Microspace diversity is routinely used in cellular base sites.

Space diversity can be used to combat short-term fading, whereas positioning satellites at different locations, not necessarily in the same orbital plane, can be used to reduce long-term fading. Space-diversity systems employ two or more antennas spaced a certain distance apart. For example, a separation of only 30 cm wavelength, which is suitable for implementation on the mobile side, is sufficient to provide a notable improvement in some mobile radio channel environments. Macrodiversity is also a form of space diversity.

Frequency-diversity systems employ two or more different carrier frequencies to transmit the same information. Statistically, the same information signal may or may not fade at the same time at the different carrier frequencies. An example of a frequency-diversity scheme includes frequency hopping and very wideband signaling. Time-diversity systems are used primarily for data transmission. The same data are sent through the radio channel as many times as necessary until the required quality of transmission is achieved. An example of this type of scheme is the well-known *automatic repeat request* (ARQ).

The improvement of any diversity scheme strongly depends on the combining techniques employed: switched combining, feedforward or feedback combining, majority vote, and so forth.

8.3 The Internet and Satellites

Satellites have offered broadband services like television or video-on-demand broadcast, wireless telephone, and electronic mail for years. Developers now place more emphasis on the speed of delivery of electronic commerce via the Internet and other telecommunication applications, which makes multimedia interactivity via the satellite very attractive. This section examines how the Internet works via the existing land-based integrated digital networks and via the satellite, as well as the technical difficulties that currently prevent rapid expansion and delivery of these services via the Internet.

8.3.1 The Internet

The Internet consists of myriad smaller, interconnected networks interoperating to exchange data between host computers in the networks. The Internet is an example of a packet-switched network. Figure 8.13 shows the basic Internet access process. The Internet user (subscriber) accesses the Internet via his or her *Internet service provider* (ISP). The ISP uses routers to connect subscribers to a dedicated-access ISDN or point of presence (POP) after concentrating subscribers' traffic.

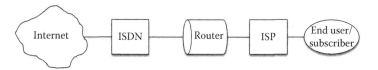

FIGURE 8.13
Basic Internet arrangement.

All networks connected to the Internet share a common suite of telecommunication protocols (standards) known as the *Transmission Control Protocol/ Internet Protocol* (TCP/IP). TCP/IP is a layered protocol suite like the open systems interconnection (OSI) protocol suite already discussed earlier in this chapter. The TCP/IP suite contains a number of protocols, some of which are shown in Figure 8.14.

The *network layer* routes packets using IP. This layer comprises other functions, for example, the Routing Information Protocol (RIP), the open shortest path first (OSPF), the Internet Control Message Protocol (ICMP), and the Address Resolution Protocol (ARP). Both OSPF and RIP are used for generating routing tables. ICMP is used for reporting errors and the status of the stations or network. ARP is used for obtaining the medium access control of a station from its IP address. IP uses a limited hierarchical address space, which has location information embedded in the structure.

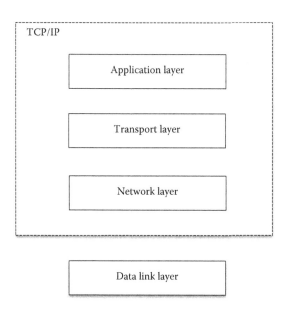

FIGURE 8.14
TCP/IP protocol suite and its relationship to the data link layer.

The *transport layer* contains TCP and ensures that a message can be received correctly by the destination. The *application layer* provides a programming interface for interprocess communication to the transport layer and application services for subscribers (or end users). Some of the many application protocols include:

File Transfer Protocol (FTP) for file transfer

Domain name service (DNS) for symbolic domain name translation to a corresponding IP network address

TELNET for text-based remote access (login)

Simple Mail Transfer Protocol (SMTP) for electronic mailing

Simple Network Management Protocol (SNMP) for network hardware and software management

Each of these protocols resides within one layer and communicates with the protocols in the adjacent layers. The overriding Internet protocol requirement is simplicity, neglecting complex functions to the edge of the network.

The *data link layer* maintains a reliable communication link between adjacent nodes. An example of this layer is the *local area network* (LAN), a communication network system that enables computer users to share computer equipment, application software, data, facsimile, voice, and video transmission.

Internet users have a wide range of new application requirements beyond the simple data transfer; these include the smart and quick interface of text, voice, and video, as well as real-time videoconferencing and multimedia applications. To meet all these demands, it is imperative that a family of robust, secure, and serviceable broadband networks is required that will be capable of providing a host of voice-, data-, and video-enhanced features and services beyond mere transport.

8.3.1.1 Security

Data security is a matter of great concern to many Internet users. Two broad types of security have been discussed in the literature and in practice: firewalls and encryption. A prominent public-key encryption method for e-mail is Pretty Good Privacy (PGP), which is based on the Rivest-Shamir-Adleman (RSA) algorithm [22]. A secured network can be built over the network by encrypting and decrypting information at both end terminals and stations. Chapter 1, Section 1.4 discusses the principles of encryption and decryption.

Firewalls attempt to provide filters for both incoming and outgoing packets in order to improve the security of the networks connected to the Internet. Examples include:

Packet filters, which use the IP address and TCP port number to decide if a packet can pass through the routers. This filter type is used at the network layer to regulate traffic flow.

Proxy servers (also called circuit-level gateways) prevent direct connection from the outside to the service on the inside.

Stateful packet inspection (also called *dynamic firewall* technology) uses a comparison basis technique, where a packet under filtering is compared with the saved state of the previously forwarded packets. If the record shows that the packet under inspection appears reasonable, then permission is granted to pass the packet to the intended destination.

Several firewall products are available in the market.

8.3.2 The Internet via Satellite

Current technology allows the Internet interaction channel to remain within the ISDN link. Satellite technologies are uniquely poised to address some of the key challenges that must be met for the Internet to continue its rapid expansion and advancement: specifically, high-speed access for rural and remote areas, distribution and delivery of media-rich content, and tight integration with existing technologies. With advances in technology and miniaturization of electronic components, bidirectional communications can be carried over satellite, eliminating the necessity of the subscribers' (end users') terrestrial Internet connections that are inherent with the existing one-way satellite systems.

Figure 8.15 shows a basic Internet access through the satellite. The Internet user (subscriber) accesses the Internet via his or her ISP, which then establishes access to an Internet point of presence via a gateway station for connection to the satellite's network access gateway (NAG). The NAG then sends the message across the proximate backbone network of the recipient's ISP, which is finally delivered to the intended recipient's address. For this type of application, channel capacity needs to be shared by different users.

Subscribers' access to a satellite using a subscriber unit (e.g., handheld, pocket-sized phones) may be the simplest feature of satellite mobile systems of the future where the subscribers will be able to access voice, facsimile, data, and paging services with acceptably high quality.

There are direct benefits for users accessing the Internet via satellite, including:

Connection through or bypassing terrestrial networks and provision for direct-to-home services via the digital video broadcasting (DVB) IP. DVB technology uses multiplexing (e.g., MPEG-2) for data packeting, but with different channel modulation. MPEG-2 is the broadcast-quality compression standard for video communication.

Provision for higher throughput and possibly asymmetric links.

Allowance for multicasting transmission, that is, sending one message to authorized users instantaneously within the network through a variety of earth stations.

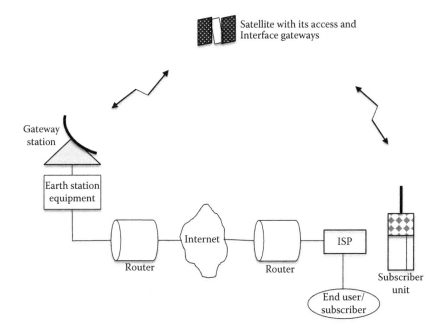

FIGURE 8.15
Internet via satellite.

Provision for faster access (or response) time and delivery of information
to any inhospitable terrain or region where there is a rapidly expand-
ing customer demand but insufficient terrestrial infrastructure.

Of course, there have been attempts to deliver IP over satellite, but the satellite technologies have focused on connection-oriented transmission protocols, which are really suited for voice traffic, rather than IP unnecessarily
squandering expensive capacity. What are the challenges ahead for effective
delivery of IP over the satellite?

8.3.3 Technical Challenges

The TCP/IP shortcomings in the typical satellite environment include
latency (degradations due to slow start), data security over the satellite
link, and optimization (window size, traffic acknowledgment for short
transmissions over high-latency channels, as well as scheduling capacity).
These shortcomings demand attention in order to provide a more efficient,
secure, and application-expansive internetworking over the satellite. The
issue of data or message security via the satellite has been discussed in
Chapter 1 and earlier in this chapter. Thus, we examine TCP/IP challenges
in this section.

8.3.3.1 Latency

Latency is a critical parameter of communication service quality, particularly for interactive communications and for many standard data protocols. TCP/IP performs poorly over high-latency or noisy channels. Bit error rates (BERs) of 10^{-7} can be acceptable in telephone environments, but this level of performance will render TCP/IP almost unusable. Geostationary satellite links are inherently high latency, and can be noisy. For example, we note in Chapter 1, Section 1.3 that if circuits are positioned in geosynchronous orbits, they suffer a transmission delay of about 119 ms between an earth terminal and the satellite, resulting in a user-to-user delay of 238 ms and an echo delay of 476 ms. Latency in voice communications becomes noticeable with a round-trip delay over 100 ms. To overcome the slow start this delay causes, an enhanced communication protocol over the satellite link must be implemented that terminates and reoriginates TCP connections at both ends of the satellite link without sacrificing the discovery process and still maintaining end-to-end standards compliance.

8.3.3.2 Optimization

Window size and scheduling capacity are optimization issues, which affect data throughput and network efficiency. Window sizing works simply by allowing a transmitter to send a number of packets before waiting for an acknowledgment from the receiver. Typically, the window size is short in comparison to the 476 ms delay of a satellite link; however, it minimizes the amount of data that would need to be retransmitted when packets are dropped. This operation dramatically slows down a satellite connection.

Scheduling capacity as well as "burstiness" of subscriber traffic over the satellite link is another area that can potentially reduce data throughput or network efficiency. There are two common ways of scheduling capacity: *connection oriented* and *connectionless*, which reflect two different philosophies of link management. Traffic burstiness determines which scheduling method is appropriate for data transmission.

The connection-oriented method assigns a fixed level of capacity prior to transmission and reassigns capacity at the completion of a transmission. What is remarkable about this method is that at least one round-trip delay is required to both allocate and de-allocate capacity. For long connections such as voice phone calls, the overhead to set up and tear down links is very small. For short connections, such as a uniform resource locator (URL) request over a network, the setup and teardown times can overwhelm any efficiency the existing protocol achieves. The traditional switched telephone network's traffic is relatively smooth and is well suited to this method. The connection-oriented method does not handle bursty traffic efficiently because bursts potentially slow down by the ceiling on capacity allocation.

Connectionless refers in part to the way in which capacity is allocated. It requires no setup and teardown time. An underlying connectionless method is a multiple-access scheme that allows for random traffic, collision detection, and an efficient method for handling collisions to avoid channel breakdown. The traffic characteristic in the data networking environments tends to arrive in bursts, which is suited to a connectionless method. The issue is how to run an efficient multiple-access scheme that addresses the issue of channel breakdown.

8.4 Summary

The unrelenting growing demand for interactive communication systems using small handheld terminals has resulted in the provision of mobile satellite system services. As demand for networks has evolved, standards organizations have worked—and still strive—to ensure compatibility of networks and networking equipment. This chapter discusses the basic architecture of mobile satellite systems as well as the backbone transmission network, channel types, channel structures, OSI and ISDN protocol reference models, and error performance standard.

Satellites excel in one-to-many communications, reaching remote areas where it is not profitable to set up a cellular or Internet network for telecommunications or data transfer. This chapter also examines the cellular network, Internet operating principles and protocols, and system security. Latency, channel congestion, capacity scheduling, and network optimization are some of the issues that require solutions to ensure reliable and secure data transmission via satellite. With rapid advances in technology, lower production costs, and insatiable appetite for the consumers' dollar—as well as the consumers' demand for rapid, secure, and reliable interactive telecommunications—the future is bright for communications via satellite.

Problems

1. You are consulting for a county, which contemplates replacing its antiquated communication systems. Besides the county's administrative center, most of its population is sparsely spread with varying terrain. What questions would you ask before deciding on your recommendations for the communication network design? Explain the rationale behind the questions.

2. Explain the difference between OSI and ISDN reference models.

3. What applications are parts of the TCP/IP protocol, and what functions do they perform?

4. Describe the protocol used on the Internet.

5. Why is a layered architecture used for network communications?

6. Explain the difference between the end user-to-network interfaces and their usefulness in the ISDN architecture.

7. What is the difference between peer protocol and virtual circuit in a reference model?

8. Is it correct to consider the interconnectivity architecture as two half gateways?

9. Design a security system that protects your system from unwarranted intrusion from other networks and mail-order marketers.

10. What is stored in a routing table?

References

1. Radiocommunication. http://www.itu.int/ITU-R/ (accessed September 10, 2012).
2. Helgert, H.J. (1991). *Integrated services digital networks.* Reading, MA: Addison-Wesley.
3. ITU. (2012). Recommendations Q-series. http://www.itu.int/ITU-R/ accessed October 1, 2012).
4. Kolawole, M.O. (2002). *Satellite communication engineering.* New York: Marcel Dekker.
5. Prasad, K.V.K.K. (2004). *Principles of digital communication systems and computer networks.* Independence, KY: Charles River Media.
6. Hanrahan, H. (2007). *Network convergence: services, applications, transport, and operations support.* New York: John Wiley.
7. Bagad, V.S. (2009). *Telecommunication switching systems and networks.* Pune: Technical Publications Pune.
8. Stallings, W. (2011). *Data and computer communications.* Englewood Cliffs, NJ: Prentice Hall.
9. Ronayne, J (1988). *The integrated services digital network: from concept to application.* London: Pitman/Wiley.
10. Wu, W.W. (1989). *Elements of digital satellite communication: channel coding and integrated services digital satellite networks.* Rockville, MD: Computer Science Press.
11. ITU-T. (1999). *ISDN user-network interfaces—interface structures and access capabilities.* ITU-T Recommendation I.412. *This recommendation is still in force.*
12. Mahoney, J. (1975). Users view of the network. *Bell Systems Technical Journal,* 54, 5.

13. ITU-T. (1988). *Public land mobile network, mobile application part and interfaces.* Recommendations Q1051–Q1063.
14. ITU-T. (1988). *Public land mobile network, interworking with ISDN and PSTN.* Recommendations Q1000–Q1032.
15. ITU-T. (1983). Study Group XVIII. Reports R3–R5 for 1981–1984.
16. Anderson, S. (1996). GSM-interoperable mobile station call processing architecture in mobile and personal satellite communications 2. *Proceedings of the European Workshop on Mobile/Personal Satcoms,* 247–269.
17. Steele, R. (1994). *Mobile radio communications.* London: Pentech Press.
18. White, G. (1995). *Mobile radio technology.* Oxford: Newnes.
19. Kangas, S., and Lahtela, K. (1989). Cell division in mobile telephone systems, in Pan-European mobile communications. *IBC Technical Services,* 91–95.
20. Kucar, A.D. (1995). Mobile radio—an overview. In *Cellular radio and personnal communications* (Rappaport, T.S., ed.). New York: IEEE Press.
21. Kolawole, M.O. (2009). *A course in telecommunication engineering.* New Delhi: S. Chand.
22. Rivest, R., Shamir, A., and Adleman, L. (1978). A method for obtaining digital signatures and public-key cryptosystems. *Communications for the ACM,* 21(2), 120–126.

Appendix A: Notations

The symbols have been chosen as carefully as possible to prevent confusion. In a few instances, the same symbol was used. When this occurs, a clear distinction is made in the meaning of each, and where it is used in the text is indicated.

Symbol	Definition
A	Aperture length in H-plane of a pyramidal horn
A_a	Solar array area required for the cells
A_c	Area covered by each radio cell
A_e	Effective aperture area
A_{eq}	Equivalent surface area of the satellite that is perpendicular to the orbit velocity
A_g	Cellular radio coverage area for circular cells
A_p	Cellular radio coverage area for polygonal cells
B	Aperture length in E-plane of a pyramidal horn (in Chapter 2) or channel bandwidth (in other chapters)
C	Average carrier power
$C_{CD/FD/TD}$	User capacity in a CDMA/FDMA/TDMA system
C_d	Coefficient of aerodynamic drag
$C_{n/x}$	Normalized total capacity for a FDMA/CDMA system
D	Aperture diameter
D_{el}, D_{az}	Aperture diameters in elevation, azimuth
F	Flux density at the receiver
F_d	Force due to aerodynamic drag
F_g	Gravitational force acting on the satellite
G	Antenna gain or generator matrix (in Chapter 6)
H	Entropy
H_Δ	Parity-check matrix
H_g	Altitude of earth station
H_o	Freezing height
I	Identity matrix
I_{di}	Power of the ith interfering signals on the downlink
I_{ui}	Power of the ith interfering signals on the uplink
$I(x_i)$	Information associated with source states x_i
K	Number of cells per channel
K_D	Decryption key
K_E	Encryption key
L	Axial length to apex to the aperture of the conical horn or rain geometric pathlength (Chapter 4)
L_1	The pyramidal horn's shortest length
L_a	Number of octets in a cell
L_c	Constraint length of the convolutional code

L_{ET}	Latitude of the earth station
L_{fs}	Free-space loss
L_p	Total path loss
L_{SAT}	Latitude of the satellite
L_T	Total antenna losses
$L(w_i \mid \mathbf{x})$	Likelihood function of data vector $\mathbf{x}$ and the parameter w_i
$M(j) \backslash i$	The set $M(j)$ with check node i excluded
M_r	Received decrypted message
M_t	Transmitted plaintext message
N	Noise power
N_c	Cell cluster size
N_f	Number of failing system components
$N(i) \backslash j$	The set $N(i)$ with variable node j excluded
N_{IU}	Total uplink noise power
N_o	Number of components comprising the system at test period
N_s	Number of surviving system components
P	Arbitrary matrix
P_r	Received power
P_s	Effective solar system power
P_T	Total radiated by transmitting source
R	Distance from the hypothetical isotropic source or spherical surface radius
R_c	Code rate
R_e	Equatorial radius of the earth
R_r	Data transmission rate
$R(t)$	System reliability in time t
R_v	Geocentric radius of earth station
S	Error syndrome
S_f	Semifocal length of an elliptic orbit
S_m	Semiminor axis of an elliptic orbit
S_p	Semiparameter of an elliptic orbit
T	System noise temperature
T_a	Antenna temperature
T_b	Bit (or burst) duration
T_c	Average time between successive channels
T_s	Symbol (or frame) duration
X	Message data
Y	Received codeword
Z	Error-prone received codeword
a	Semimajor axis of an ellipse (Chapter 2), unpolarized rain attenuation coefficient (Chapter 4), or side of a polygon (Chapter 8)
a_c	Attenuation coefficient for circularly polarized raindrop
a_d	Acceleration due to aerodynamic drag
a_f	Loss factor for solar array cells
a_p	Power attenuation
b	Message length

b	Unpolarized rain attenuation coefficient
b_c	Attenuation coefficient for circularly polarized raindrop
c	Speed of light
$cov[\mathbf{x}]$	Covariance matrix of vector $\mathbf{x}$
d_c	Channel density
$d_{H,min}$	Hamming minimum distance
d_i	Frequency band reuse distance
e	Eccentricity of an ellipse
e_k	Number of correctable errors per codeword in the BCH coding techniques
e_l	Electrical losses of solar cells
f_1	Local oscillator frequency
f_c	Incoming signal frequency
g	Acceleration due to gravity at the surface of the earth
h	Number of octets in the header
h_o	Altitude above the subsatellite point on the earth terminal
k	Boltzmann's constant
k_e	Constant used to emphasize the type of aperture
k_s	Solar constant
l	Number of accesses or users
m	Reference bursts
m_k	Length of shift register
m_s	Satellite mass
n	Block length
n_c	Number of cells in a designated radio cell area
n_{ch}	Number of radio channels in a cell
n_p	Number of bursts, accesses, or traffic stations to the frame
$p(x)$	Probability of x
$p(x \mid y)$	Probability of transmitting x given that y was received
$p(y)$	Probability of y
q	Length of burst errors
r	Radius distance focus to the orbit path
r_b	Input digital bit rate of the user (bit/s)
r_c	Average radius of a radio cell
r_r	Rain rate
sgn	Signum
t_d	Transmission delay between an earth terminal and satellite
t_e	Rotation period of the earth
t_p	Amount of time an earth station is required to communicate with an orbiting satellite as it passes overhead
t_s	Orbit period
v	Velocity of the satellite with respect to the atmosphere
w_s	Swath width
$\hat{x}_{i,MAP}$	Maximum a posteriori (MAP)
α	Apex angle for global coverage
α_H	Magnetic heading of the antenna

$\alpha_{i,j}^{n}$	For nth iteration, the message sent from variable node j to check node i
β	State probability
$\beta_{i,j}^{n}$	For nth iteration, the message sent from check node i to variable node j
Δ	Difference in longitude between the earth station and the satellite
Δ_{α}	Deviation angle between true north and north magnetic poles
Δf_{ch}	Total bandwidth assigned to the service
Δf_{u}	Bandwidth allocated to each user
ε	Error vector
γ	Central angle
ζ	Channel gain
η	Antenna efficiency (Chapter 2) or throughput (Chapter 5)
η_{m}	Solar cells' conversion efficiency
λ	Wavelength in free space
λ_{r}	Average failure rate
θ	Earth station elevation angle
Θ	Deviation angle from the electric axis of the antenna
θ_{BW}	3 dB antenna beamwidth
θ_{E}	3 dB antenna beamwidth in E-plane
θ_{H}	3 dB antenna beamwidth in H-plane
θ_{t}	True elevation angle
ρ_{a}	Atmospheric density
ω	Angular frequency

Appendix B: Glossary of Terms

Antenna: The interface between a free-space electromagnetic wave and a guided wave. It can also be described as the input and output interface of a spacecraft.

Antenna illumination: The distribution of radio energy across an antenna or reflector surface. Illumination affects antenna gain. Distribution of the radio energy depends on the feed horn.

Aperture: The surface area of an antenna that is exposed to a radiofrequency (RF) signal.

Apogee: The highest point along an elliptical orbit where the orbit is farthest from earth.

Attenuation: The power loss between one transmitting source and the receiver due to path losses.

Bandwidth: The range of frequencies required to represent the information contained in a signal.

Bearer service: A type of telecommunication service that provides the capacity for the transmission of signals between the user and network interface.

Bit error rate (BER): An empirical record of a system's actual bit error performance.

Cell: A short block of information, usually of fixed length, that is identified with a label.

Channel: A specific portion of the information carrying capacity of a network interface, and specified by a specific transmission rate.

Channel structure: The structure that defines the maximum digital carrying capacity in terms of bit rates across a network interface.

Circuit switching: The method of transferring information in which switching and transmission functions are achieved by permanently allocating a number of channels or bandwidth between the connections.

Codec: A contraction of *code* and *decode*.

Code division multiple access: A multiple-access scheme where stations use spread-spectrum modulations and orthogonal codes to avoid interfering with one another.

Common channel signaling network: The network that provides physical and transmission capacity for the transfer of connection control signals between components of interexchange network.

Communication satellite: An electronic relay station orbiting in space that picks up messages transmitted from the ground and retransmits them to a distant location.

Constellation: A collection of similar satellites designed to provide multiple coverage or multiple redundancy.

Cryptographic key management: The generation, distribution, recognition, and reception of the cryptographic keys (passwords).

Cryptography: A science that comprises encryption, decryption, and a cryptographic key management unit.

Decoding: The process required to reconstruct the data bit sequence encoded onto the carrier.

Decryption: The unlocking of the locked (decrypted) message; the reverse of encryption.

Demodulation: A process by which the original signal is recovered from a modulated carrier; the reverse of modulation.

Delay: Time a signal takes to go from the sending station through the satellite to the receiving station.

Downlink: The space-to-earth telecommunication pathway.

Drag: The friction between a moving vehicle through the earth's atmosphere opposing the vehicle's forward motion.

Earth station: A terrestrial terminal that provides a means of transmitting the modulated radiofrequency (RF) carrier to the satellite within the uplink frequency spectrum and receiving the RF carrier from the satellite with the downlink frequency spectrum. It is the vital link between satellite and communication users.

Encryption (enciphering): The process of converting messages, information, or data into a form unreadable by anyone except the intended recipient.

Feed: The radiofrequency (RF) input/output device.

Footprint: An area on earth illuminated by a satellite antenna. Its shape and size depend on a number of factors, including the antenna design and the satellite's angle of elevation.

Forward error correction: A transmission error detection and correction technique that, in addition to detecting errors, also corrects errors at the receive end of the link.

Frequency division multiple access: A technique by which the transponder bandwidth is shared in separate frequency slots.

Frequency reuse: Using the same frequency band several times in such a way as to increase the overall capacity of a network without increasing the allocated bandwidth.

Functional group: A set of functions required for carrying out tasks in cooperation with another functional element or group.

Gain: A measure of amplifier power expressed in watts (W) or decibels (dB).

Gateway: A special switch used to link part of a network.

Geostationary satellite (GEO): A satellite placed in the geosynchronous orbit in the equatorial plane. Although it appears not to be moving when seen from the earth, its velocity in space equates to 3.076 km/s (11,071.9 km/h).

Gigahertz (GHz): One billion (10^9) cycles per second.

High elliptical orbiting satellite (HEO): A specialized orbit in which a satellite continuously swings very close to the earth, loops out into space, and then repeats its swing-by. It is an elliptical orbit approximately 18,000–35,000 km above the earth's surface, not necessarily above the equator.

Horn: An open-ended waveguide designed to radiate maximum power in one direction. It could also be part of a feed.

INMARSAT: International Maritime Satellite Organization. Primarily operates telecommunication services for ships at sea.

Internet (or internetworking): The interconnection of multiple networks into a single, large virtual network.

Intersatellite link (ISL): The link between two satellites in the same orbit.

ISDN: Integrated services digital network. A telecommunications network that provides end-to-end digital connections to end users and other network facilities.

JPEG: Joint Photographic Experts Group. The standard used for compressing and decompressing continuous-tone (color or grayscale) images. The ISO/IEC JTC1/SC29 Working Group 10 sets this standard.

Ka-band: Frequency range from 18 to 31 GHz.

Kbit/s: Data transmission speed of 1000 bits/s.

Ku-band: Frequency range from 10.9 to 17 GHz.

L-band: Frequency range from 0.5 to 1.5 GHz. Also used to refer to the 950 to 1450 MHz range used for cellular radio communications.

Low earth orbiting satellite (LEO): A satellite that orbits the earth in grids, stretching approximately 160–1600 km above the earth's surface.

MAP: Maximum a posteriori probability (Chapter 6)—an optimal estimate of an unobserved quantity on the basis of empirical data; or, Management application process (Chapter 8)—the network application that defines services for signaling among several mobile switching centers.

Middle earth orbiting satellite (MEO): A compromise satellite between the lower orbits and the geosynchronous orbits, circular, and orbits approximately 8000–18,000 km above the earth's surface, not necessarily above the equator.

Mobile satellite system (MSS): A radio communication service between mobile earth stations and one or more space stations, between mobile earth stations by means of one or more space stations, or between space stations.

Modem: A converter that performs *modulation* and *demodulation* of transmitted data.

MLE: Maximum likelihood estimation. An estimate that seeks the probability distribution that makes the observed data most likely.

MPEG: Motion Picture Expert Group. The standard used for compressing motion pictures and multimedia, in contrast to JPEG. The ISO/IEC JTC1/SC29 Working Group 11 sets this standard.

Multicasting: Sending one message to specific groups of users within a network.

Multiplexers: Devices that are part of a switching matrix in the satellite communication subsystem that distribute, connect, and combine signals for the amplifiers and antennas.

Multiplexing: The ability to access a single transponder circuit by multiple users.

Network: A collection of links and nodes that provides connections between two or more defined points to facilitate communication between them.

Omnidirectional antenna: An antenna that radiates uniformly in all directions.

Orbit: A path, relative to a specified frame of reference, described by the center of mass of a satellite in space, subject to gravitational attraction.

Packet switching: A data transmission method that divides messages into standard-sized packets for greater efficiency of routing and transport through a network.

Perigee: The lowest point along an elliptical orbit, where the satellite is closest to the earth.

Protocol: A set of rules specified to ensure that a disciplined and accurate transmission and exchange of information occur across a communication channel.

Protocol reference model: A framework for the hierarchical structuring of functions in a system and its interaction with another system.

PSTN: Public switched telephone network. A telecommunications network that provides end-to-end telephone connections to end users and other network facilities.

Receiver: An electronic device that enables the desired modulated signal to be separated from all the other signals coming from an antenna.

Router: Same as gateway.

Satellite: A sophisticated electronic communication system positioned in an orbital plane above the earth. Alternatively, a body that revolves around another body of preponderant mass and that has a motion primarily and permanently determined by the force of attraction of this body.

Scrambler: A device that electronically alters a modulated signal so that only those recipients equipped with special decoders can pick up and interpret an undefiled version of the original signal.

Signaling system: A set of protocols for the exchange of signaling information, for example, SS7.

Signaling transfer point (S/T): A signaling point whose primary function is to transfer signaling messages from one signaling link to another.

Single channel per carrier: A multiple-access technique in which each distinct user channel is allocated a specific carrier instead of a number of channels being multiplexed onto a single carrier.

Subscriber access network: The functional part of ISDN that lies between the end user or subscriber and the common channel signaling network (CCSN) and interexchange network (IEN).

Teleservice: A type of telecommunication service that provides both bearer capabilities and higher-layer capabilities for communication between users.

Terminal equipment (TE): A functional group that includes functions necessary for protocol handling, interfacing, maintenance, and connection with other equipment.

Time division multiple access: A form of multiple access where a single carrier is shared by many users. Signals from earth stations reaching the satellite consecutively are processed in time segments without overlapping.

Transmitter: An electronic device that broadcasts modulated signals toward one or many distant receivers.

Transponder: An electronic device carried onboard a communication satellite that picks a signal on one frequency, amplifies it, and retransmits it on another frequency.

Uplink: The earth-to-space telecommunication pathway.

User-to-network interface: The point of demarcation between the terminal equipment and the network termination.

X.21: A recommendation by ITU-T that specifies the interface between a data terminal equipment (DTE) and a data-circuit terminal equipment (DCE) for synchronous operation over public circuit-switched data networks.

X.25: A recommendation by ITU-T that specifies the interface between a data terminal equipment (DTE) and a data-circuit terminal equipment (DCE) for operation in virtual circuit mode over a packet-switched network.

Index

A

access, 159
accumulate-repeat-accumulate (ARA) code, 193
actual equalizer output, 106
additive white Gaussian noise (AWGN)
    error-free performance, 95
    likelihood functions, 198
    multiple-access methods, capacity comparison, 170
Address Resolution Protocol (ARP), 251
Advanced Encryption Standard (AES), 10
AES, *see* Advanced Encryption Standard (AES)
algorithms, encryption, 6–9
amplitude shift keying (ASK), 85, *see also* Modulation
analysis
    availability, 44–48
    satellite system engineering, 41–42
AND logic
    cyclic codes, 190
    likelihood functions, 200
    linear block codes, 183
Anik satellite, 2
antennas
    defined, 263
    helical type, 59–60
    horns, 50–53
    illumination, 263
    overview, 49–50
    phased-array type, 57–59
    reflector/lens system, 53–57
    selection criteria, 44
    tracking, 114–118
antipodal signals, 85
aperture, 263
apex angle, 31
apogee, 28, 263, *see also* Perigee

a posterior probability (APP), 199
APP, *see* A posterior probability (APP)
application layer
    Internet, 252
    open systems interconnection, 239
ARA, *see* Accumulate-repeat-accumulate (ARA) code
architecture
    call setup, 243
    cell division, 246–248
    cell size, 243–246
    cellular mobile systems, 242–250
    channel structures, 230, 235–237
    channel type, 230–235
    error performance standard, 237
    fading and diversity schemes, 249–250
    frequency reuse, 248–249
    integrated services digital network, 230–242
    OSI reference model, 238–239
    overview, 230–233
    protocol reference model, 239–242
ARP, *see* Address Resolution Protocol (ARP)
ARQ, *see* Automatic repeat request (ARQ)
ASK, *see* Amplitude shift keying (ASK)
asynchronous channels, 236–237
atomic clocks, 71–72
attenuation, 263
Aussat satellite, 2
automatic repeat request (ARQ), 250
autotracking system, 151
availability analysis, 44–48
average cross-correlation, 161
AWGN, *see* Additive white Gaussian noise (AWGN)
axisymmetric reflectors, 56
azimuth angle
    geometric distances, 35, 37
    visibility of satellites, 38

Printed in the United States
by Baker & Taylor Publisher Services